Mathematische Optimierung und Wirtschaftsmathematik | Mathematical Optimization and Economathematics

Reihe herausgegeben von
R. Werner, Augsburg, Deutschland
T. Harks, Augsburg, Deutschland
V. Shikhman, Chemnitz, Deutschland

T0206395

In der Reihe werden Arbeiten zu aktuellen Themen der mathematischen Optimierung und der Wirtschaftsmathematik publiziert. Hierbei werden sowohl Themen aus Grundlagen, Theorie und Anwendung der Wirtschafts-, Finanz- und Versicherungsmathematik als auch der Optimierung und des Operations Research behandelt. Die Publikationen sollen insbesondere neue Impulse für weitergehende Forschungsfragen liefern, oder auch zu offenen Problemstellungen aus der Anwendung Lösungsansätze anbieten. Die Reihe leistet damit einen Beitrag der Bündelung der Forschung der Optimierung und der Wirtschaftsmathematik und der sich ergebenden Perspektiven.

More information about this series at http://www.springer.com/series/15822

Manuela Spangler

Modelling German Covered Bonds

 Springer Spektrum

Manuela Spangler
Institute of Mathematics,
Business Mathematics
University of Augsburg
Augsburg, Germany

Zugl.: Dissertation, Universität Augsburg, 2018

Erstgutachter: Prof. Dr. Ralf Werner
Zweitgutachter: Prof. Dr. Marco Wilkens
Tag der mündlichen Prüfung: 19.1.2018

The views and opinions expressed in this work are those of the author only, and not those of her employer, MEAG MUNICH ERGO AssetManagement GmbH, its subsidiaries or affiliate companies. Nothing in this work is or should be taken as information regarding, or a description of, any operations, risk appraisal or consulting business of MEAG MUNICH ERGO AssetManagement GmbH.

ISSN 2523-7926   ISSN 2523-7934   (electronic)
Mathematische Optimierung und Wirtschaftsmathematik | Mathematical Optimization and Economathematics
ISBN 978-3-658-23914-5   ISBN 978-3-658-23915-2   (eBook)
https://doi.org/10.1007/978-3-658-23915-2

Library of Congress Control Number: 2018958362

Springer Spektrum
© Springer Fachmedien Wiesbaden GmbH, part of Springer Nature 2018

This Springer Spektrum imprint is published by the registered company Springer Fachmedien Wiesbaden GmbH part of Springer Nature
The registered company address is: Abraham-Lincoln-Str. 46, 65189 Wiesbaden, Germany

Acknowledgements

This dissertation is the product of my research as an external doctoral student at the Faculty of Mathematics, Natural Sciences, and Materials Engineering at the University of Augsburg. It has been inspired by the experience gained from various years of working as a quantitative risk analyst in the finance sector.

At this point I want to express my gratitude to several people who supported me during this project and contributed to the success of the work.

First of all, I would like to thank my supervisor Prof. Dr. Ralf Werner for his continuous assistance and support over the years. He encouraged my research and has been a tremendous mentor for me. With his knowledge and experience, he always gave me inspiration and advice when needed. Without his guidance, this dissertation would not have been possible.

I am also indebted to Prof. Dr. Marco Wilkens for becoming my co-referee, and for his valuable academic advice.

Special thanks go to Max Hughes, Jan Natolski, Selma Uhlig, Tje Lin Chung and Florian Arnecke for their helpful academic comments, and to Werner Neumeier, for many fruitful discussions on the practitioner's perspective.

Moreover, I am very thankful to Peter Schenk and Barbara Götz (both from MEAG MUNICH ERGO AssetManagement GmbH), Marco Hauck, Pierre Joos and Christoph Winter (all from Allianz Deutschland AG) and Klaus Böcker (Deutsche Pfandbriefbank AG), who always granted me the necessary flexibility to combine my work life and my dissertation in an efficient way.

Last but not least, I would like to thank Stefan and Leila, my parents and my friends for their continuous encouragement and support, especially during the final phase of my dissertation.

Manuela Spangler

Contents

List of Figures

List of Tables

Abstract

With a long track record going back to the eighteenth century and no single case of default until today, the German Pfandbrief has an undisputed benchmark status in the covered bond market. It survived the recent financial and sovereign crises comparably unharmed and has proved to be a reliable and stable funding instrument also in times of market distress. Nevertheless, as the past with several Pfandbrief bank bailouts has shown, Pfandbriefe cannot be considered to be completely risk-free, despite their high level of protection. To adequately model the risks arising from a Pfandbrief investment, it is not sufficient to consider only the creditworthiness of the issuer. Product-specific features and the quality of the cover pool also need to be taken into account.

In this work we develop a multi-period simulation-based Pfandbrief model which accounts for the product's most important characteristics and adequately reflects its main risks. The model distinguishes between bank and cover pool default and considers two different default triggering events: overindebtedness and illiquidity. Both default events are influenced by the market environment, which is represented through the stochastic dynamics of the short rate and the creditworthiness of the bank's risky assets, and the resulting liquidation payments take into account the Pfandbrief-specific priority of payments. The asset liability management in our model is dynamic and considers funding and reinvestment strategies as well as the maintenance of overcollateralization according to the legal requirements.

The model's primary outputs are Pfandbrief default statistics. Simulation results obtained from an exemplary Mortgage Pfandbrief calibration with typical asset liability mismatches capture the main expected behaviour patterns of Pfandbriefe. Due to its modular setup, our model provides a flexible framework for structural analyses and can be easily extended for tailor-made investigations. Potential areas of applications include but are not limited to the comparison of different Pfandbrief risk profiles and studies in the context of current policy debates such as the introduction of extendible Pfandbrief maturities.

1 Introduction

A Pfandbrief is a covered bond issued under German Pfandbrief legislation. The main feature of covered bonds is their dual protection mechanism with full recourse to the issuer and, in case of issuer insolvency, a preferential claim on a dedicated set of assets, the cover pool. Covered bonds provide the issuer with cost-efficient long-term funding, mostly for mortgage and public-sector loans. For the investor they offer an attractive investment opportunity with high credit quality and privileged treatment in various areas of EU financial market regulation.[1] Covered bonds are primarily bought by institutional investors such as banks, asset managers, insurers, pension and investment funds and central banks. They play an important role in Europe's capital markets[2] and in the context of financial stability. According to the European Covered Bond Council (ECBC), the total outstanding covered bond volume was EUR 2,498 bn at the end of 2015, with new issuances of EUR 540 bn in that year (ECBC [62]).[3] Active markets can be found in more than 20 European countries and there are also non-European countries with notable covered bond issuance, including Australia, Canada, New Zealand, Singapore and South Korea. The number of countries with a covered bond legislation in place is expected to grow further as there are several countries which are planning to update or adopt covered bond legislation. There is, however, no common covered bond legislation in place. Instead, each jurisdiction relies on its own regulation.

This work is dedicated to the German Pfandbrief which, based on 2015 ECBC data, has the highest share in today's covered bond markets with an outstanding volume of EUR 384 bn, followed by Denmark (EUR 383 bn),

---

[1] Covered bonds benefit from lower risk weights under the Capital Requirement Regulation (CRR), lower spread-risk charges under Solvency II, the categorization as highly liquid asset class under the Liquidity Coverage Ratio (LCR) and exemption from bail-in under the Bank Recovery and Resolution Directive (BRRD).

[2] According to Grossmann and Stöcker [80], the covered bond market constitutes the most important segment of privately issued bonds on European capital markets.

[3] At the time of writing, 2016 covered bond statistics had not yet been published by the European Covered Bond Council (ECBC).

© Springer Fachmedien Wiesbaden GmbH, part of Springer Nature 2018
M. Spangler, *Modelling German Covered Bonds*, Mathematische Optimierung und Wirtschaftsmathematik | Mathematical Optimization and Economathematics, https://doi.org/10.1007/978-3-658-23915-2_1

France (EUR 323 bn) and Spain (EUR 281 bn). Given its long track record going back to the eighteenth century and no single case of default until today, the Pfandbrief has an undisputed benchmark status. It survived the recent financial and sovereign crises comparably unharmed and proved to be a reliable and stable funding instrument also in times of market distress. This is, amongst others, due to the strong legislative framework for Pfandbrief issuance and the large systemic support for this product. Nevertheless, despite the high level of protection, Pfandbriefe cannot be considered to be risk-free, cf. Spangler and Werner [145]. As the past with several Pfandbrief bank bailouts has shown, the risk of failing issuers cannot be neglected, which means that structural Pfandbrief features and individual cover pool characteristics also need to be considered. With the recent introduction of bank resolution systems, government support for failing banks is likely to decrease further and cannot be relied upon any more in future.

Notwithstanding their long history and high importance in European capital markets, there is surprisingly little academic literature on the quantitative risk modelling of Pfandbriefe. Notable exceptions are the one-period Pfandbrief model introduced by Sünderhauf [148], a structural comparison of Pfandbriefe, CDOs and MBS with focus on credit risk implications by Rudolf and Saunders [139], an analysis of Pfandbrief spread time series and default risk premia by Siewert and Vonhoff [144] and the one-period approach by Tasche [149], which focuses on the impact of asset encumbrance (i.e. the reservation of certain assets for specific creditors) by the cover pool. To our best knowledge there is so far no multi-period Pfandbrief model which allows for structural analyses in the context of product-specific features and risk profiles.

The purpose of this work is to close this gap by developing a new Pfandbrief model which accounts for the product's most important characteristics and adequately reflects its main risks. A multi-period simulation-based framework is our method of choice. We use a discrete-time setup in which all terms can be easily derived by recursive computation. Similar cash flow approaches are commonly used by insurance companies, in the context of market-consistent valuation of insurance liabilities, see DAV [110]. Our model, which is inspired by Sünderhauf [148] and Liang et al [114], distinguishes between bank and cover pool default and takes into consideration two different default triggering events: overindebtedness and illiquidity. While overindebtedness is caused by a deterioration of asset quality, illiquidity stems from the inability to raise enough funding to fulfil payment obligations. Apart from model calibration, our main challenge is to find a trade-off between realistic modelling assumptions and reasonable model complexity. To a certain degree, our model can be considered to be a multi-period modified extension of Sünderhauf's one-period model. Our balance sheet is, however, more granular and due to simpler stochastic processes we obtain analytical formulas for asset pricing.

The model's primary outputs are Pfandbrief default statistics. While extensive analyses in the context of structural product features are out of scope for this work, the model has a modular setup which allows for straightforward modifications and extensions to handle a broad range of tailor-made investigations. Simulation results obtained from an exemplary Mortgage Pfandbrief calibration with typical asset liability mismatches capture the main expected behaviour patterns of Pfandbriefe.

The outline of this work is as follows. In Chapter 2 we start with an introduction to the legal framework and the risks inherent in a Pfandbrief from an investor's point of view. We also identify the most important product features which should be incorporated in a realistic Pfandbrief model. Chapter 3 then gives an overview of the existing credit risk literature with particular focus on Pfandbrief modelling. As it turns out that none of the models found in the literature fulfils all our requirements, a new Pfandbrief model is proposed in Chapter 4. The next chapter, Chapter 5, is dedicated to an exemplary model calibration. In Chapter 6 we discuss the simulation results obtained from this calibration and illustrate the influence of important model parameters on liability default statistics by means of sensitivity analyses. Finally, Chapter 7 concludes and gives an outlook on potential model applications and possible areas of future research.

2 Pfandbrief Characteristics

In this chapter we start with an introduction to the main characteristics
of Pfandbriefe and discuss the risks associated with the product from an
investor's point of view. Our focus is on the two most common Pfandbrief
types in Germany, the Mortgage Pfandbrief and the Public Pfandbrief.
In Section 2.3 we then derive implications for Pfandbrief modelling by
identifying the most important features which should be accounted for by a
realistic Pfandbrief model. For a structured in-depth analysis of the legal
framework and the risks inherent in a Pfandbrief, we refer to Spangler and
Werner [145] and the extensive collection of Pfandbrief-related literature
listed therein. Sections 2.1 and 2.2 are almost entirely (and to a large
extent literally) taken from this book. Where necessary, changes to the legal
framework as comprised in the most recent legislative materials, see vdp
[157], were incorporated.

2.1 Main Product Features

A Pfandbrief is a covered bond issued under German Pfandbrief legislation.
Depending on the collateral backing the issuance, different Pfandbrief types
are distinguished. Pfandbriefe can be secured by claims against public sector
debtors (*Public Pfandbrief*) or by mortgages on real estate properties (*Mortgage Pfandbrief*), ships (*Ship Pfandbrief*) or aircraft (*Aircraft Pfandbrief*).
A bank which holds a license for Pfandbrief issuance is called *Pfandbrief
bank*.

In Germany, Pfandbrief issuance is based on a legislative framework. The
Pfandbrief Act (PfandBG), which was introduced in 2005 and amended in
2009, 2010, 2013, 2014 and 2015, establishes the provisions governing the
issuance and collateralization of Pfandbriefe. It specifies the conditions under
which a bank is granted a Pfandbrief license and the safety standards which
must be fulfilled. It also sets out rules in the context of risk management,
supervision and transparency requirements. The PfandBG supersedes the

© Springer Fachmedien Wiesbaden GmbH, part of Springer Nature 2018
M. Spangler, *Modelling German Covered Bonds*, Mathematische Optimierung
und Wirtschaftsmathematik I Mathematical Optimization and Economathematics,
https://doi.org/10.1007/978-3-658-23915-2_2

general bankruptcy regulation and is supplemented by several statutory orders which are published by the German Financial Supervisory Authority (BaFin): the Net Present Value Regulation (PfandBarWertV), the Regulation on the Determination of the Mortgage Lending Value (BelWertV), the Regulation on the Determination of the Mortgage Lending Value of Ships and Ships under Construction (SchiffsBelWertV), the Regulation on the Determination of the Mortgage Lending Value of Aircraft (FlugBelWertV), the Cover Register Statutory Order (DeckRegV) and the Funding Register Statutory Order (RefiRegV). In the following, we do not explicitly deal with the legislative materials but focus on the main product features arising from these provisions as depicted in Figure 2.1.

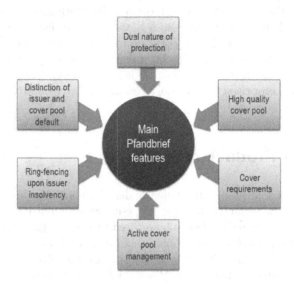

Figure 2.1: Main features of the German Pfandbrief.

Dual nature of protection. The Pfandbrief's most distinguishing feature is its dual protection mechanism. In the first place, the Pfandbrief holder has full recourse to the issuer, and in case of issuer insolvency he also has a preferential claim on a dedicated set of assets, the cover pool. Cover pool assets (which can be either claims against public sector debtors or mortgages on real estate properties, ships or aircraft) remain on the issuer's balance sheet and are registered in the cover register. For each Pfandbrief type there is a separate cover register, and cover pool assets are made identifiable by means of entry in the respective register. There is no cross-collateralization, i.e. if the bank has issued more than one Pfandbrief type, the Pfandbrief holder's preferential claim only refers to one particular cover pool. Figure 2.2 illustrates the dual protection mechanism.

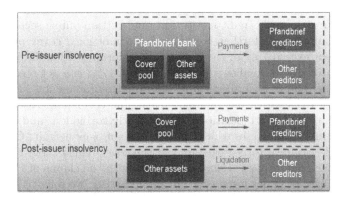

Figure 2.2: The Pfandbrief's dual protection mechanism (Source: Spangler and Werner [145], p. 2).

High quality cover pool. There are strict eligibility criteria for the inclusion of assets in the cover pool. For *ordinary cover assets* these criteria primarily depend on the Pfandbrief type. Mortgage Pfandbriefe must be backed by loans which are secured by real estate liens (both commercial and residential mortgages are allowed) and located in member states of the European Union (EU), in the European Economic Area (EEA), in Switzerland, the United States, Japan, Canada, Australia, New Zealand or Singapore.[1] Each loan can serve as cover up to 60% of its mortgage lending value.[2] Public sector cover pools, on the contrary, must consist of claims against public sector debtors (central governments and sub-sovereigns) in EU and EEA member states and, under certain additional rating restrictions, in Switzerland, the

---

[1] A limit of 10% applies for mortgage loans from outside the European Union.

[2] The mortgage lending value is defined in the BelWertV and reflects the long-term sustainable value of the property. As opposed to the market value, it ignores speculative aspects and does not depend on the business-cycle.

United States, Japan or Canada.[3] For a detailed list of the requirements
for ordinary cover assets, we refer to the legislative materials in vdp [157].
In order to increase the liquidity of the cover pool without changing its
basic characteristics, *further cover assets* can be added. For Mortgage and
Public Pfandbriefe, these include claims against suitable credit institutions,
which are eligible to up to 10% of outstanding Pfandbriefe with a limit of
2% for each single credit institution. In addition, mortgage cover pools can
also contain certain other liquid assets up to a total limit of 10% (including
claims against suitable credit institutions and certain public sector bonds).
The total amount of further cover assets in mortgage cover pools may,
however, not exceed 20% of the outstanding Mortgage Pfandbriefe. *Cover
pool derivatives* are also allowed up to a limit of 12% on a net present value
basis, given that they fulfil certain additional requirements. They have
to replicate risks which are inherent in eligible cover pool assets, which
implies that open short positions are explicitly forbidden, and must not
contain termination clauses which apply in case of bank default. Cover pool
derivatives also have to be registered in the cover register and the claims
of derivative counterparties rank pari passu with the claims of Pfandbrief
holders in case of issuer insolvency. In the following, we refer to registered
cover pool assets and derivatives as *cover pool*, and to Pfandbrief holders and
counterparties from cover pool derivatives as *privileged creditors* or *preferred
creditors*.

Cover requirements. The PfandBG specifies several cover requirements to
be maintained by the Pfandbrief bank at all times and separately for each
Pfandbrief type. First of all, the nominal of the cover pool must be equal to
or exceed the nominal of outstanding Pfandbriefe (*nominal cover*). Second,
the net present value of the cover pool has to exceed the net present value
of outstanding Pfandbriefe by at least 2% (*excess cover*). This excess cover
still needs to be given under specified interest rate and currency stresses
(*excess cover under stress*). Claims from cover pool derivatives must be
covered, too. According to the PfandBarWertV, the net present values have
to be calculated by discounting future cash flows with the currency-specific
swap curve and converting them into euros at the current exchange rate.
For derivatives, market prices have to be taken. With respect to the stress
scenarios, there are three potential methods which can be used: a static
approach, a dynamic approach and a method based on the bank's internal

---

[3] As in the case of Mortgage Pfandbriefe, a limit of 10% applies for claims from outside
the European Union.

risk model. Once the excess cover or the excess cover under stress is not given any more, additional highly liquid assets need to be posted to the cover pool. Forth, the maximum cumulative net cash outflow occurring within the next 180 days must be covered by highly liquid assets (*180-day liquidity buffer*), and the cumulative difference between cash inflows from the cover pool and scheduled liability payments to preferred creditors needs to be determined with daily granularity. To ensure that the cover requirements are fulfilled, the Pfandbrief bank performs so-called *matching cover calculations*. While the nominal cover, the excess cover and the 180-day liquidity buffer have to be monitored on a daily basis, the excess cover under stress needs to be checked on a weekly basis. Since the 2014 amendment of the PfandBG, BaFin has the competence to increase the required overcollateralization for individual Pfandbrief programmes if deemed necessary (*cover add-on*), which facilitates a reaction to unfavourable issuer- or cover pool-specific developments. This additional overcollateralization (OC) needs to be held until the prerequisites which originally triggered the add-on are not fulfilled anymore, but for at least three months. The safety buffer resulting from the legal cover requirements is also referred to as *mandatory overcollateralization*.

Active cover pool management. Pfandbrief cover pools need active management. To ensure that cover requirements are fulfilled at all times, the issuer may have to replace assets in case of asset repayments, defaults or prepayments or to add liquid assets to the cover pool. It also might be necessary to post additional assets before new Pfandbriefe can be issued, or the bank may decide to replace non-performing assets in the cover pool for reputational or marketing purposes. In practice, Pfandbrief banks tend to maintain overcollateralization levels which exceed the legal requirements. This *voluntary overcollateralization* is often driven by rating agency requirements and needed to obtain a certain target rating. It provides an additional buffer for the Pfandbrief investor. To ensure transparency, Pfandbrief banks are obliged to publish detailed information on their cover pools and their outstanding Pfandbriefe on a quarterly basis (§28 PfandBG). In the following, we refer to the associated reports as *§28 Pfandbrief statistics*.

Ring-fencing upon issuer insolvency. In case of issuer insolvency, cover pools are exempt from the bank's general insolvency proceedings. To ensure the preferential claim of the privileged creditors, the cover pools are separated from the general insolvency estate and managed by a cover pool administrator.

The ring-fenced cover pool and the corresponding outstanding Pfandbriefe continue as a Pfandbrief bank with limited business activity, the purpose of which is the full and timely repayment of the privileged creditors. In case the Pfandbrief bank has issued more than one Pfandbrief type, there would be more than one Pfandbrief bank with limited business activity, each having a separate fate. This separation principle, which is also referred to as *ring-fencing*, is illustrated by Figure 2.3.

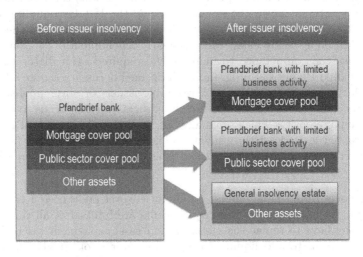

Figure 2.3: Ring-fencing upon issuer insolvency (Source: Spangler and Werner [145], p. 17).

The cover pool administrator manages the cover pool and is responsible for timely payments to privileged creditors. He can use cover pool payments to service outstanding Pfandbriefe according to their contractual terms and raise liquidity by refinancing activities such as central bank funding, take up refinancing loans or issue new Pfandbriefe. Furthermore, he has the possibility to raise funds by selling cover assets, which is facilitated by the fact that he can use an existing or newly created refinancing register. The cover pool administrator is, however, not forced to manage the cover pool on his own. With the written consent of BaFin, he can transfer the whole cover pool or parts of it as a package together with the outstanding Pfandbriefe to another Pfandbrief bank. In case of a partial cover pool transfer, it needs to be ensured that the remaining part of the cover pool still fulfils the cover requirements. The cover pool administrator can also hold the corresponding (part of the) cover pool in a fiduciary capacity for the other Pfandbrief bank. In any of these three cases scheduled payments to privileged creditors continue, which implies that issuer default itself does not necessarily have an impact on Pfandbrief payments. Once all preferred creditors have been fully repaid, the remaining assets and derivatives in the cover pool must be released to the issuer's general insolvency estate.

Distinction of issuer and cover pool default. Upon issuer insolvency, payments to privileged creditors are made out of the proceeds of the cover pool. In case the cover pool itself becomes insolvent at the time of or after issuer default, insolvency proceedings with respect to the cover pool are initiated. The cover pool is then liquidated and its liquidation proceeds are used to repay the preferred creditors. If these proceeds are not sufficient, preferred creditors also have a claim against the issuer's general insolvency estate, which ranks pari passu with the bank's other creditors. Figure 2.4 illustrates the situation post-issuer insolvency. As opposed to the event of issuer default, cover pool default has an immediate impact on Pfandbrief payments as it triggers an early repayment of the outstanding Pfandbriefe. The PfandBG does not define specific criteria for the event of cover pool default. It requires the cover pool administrator to monitor the intrinsic value of the cover pool with respect to the cover requirements on a regular basis, but it does not mention any particular consequences resulting from a breach of these requirements. This is as opposed to some other covered bond frameworks where cover pool default is explicitly defined. UK structured covered bonds, for example, specify cover pool default by means of an amortization test. This test is performed once a credit event with respect to the covered bond issuer has occurred and verifies whether the cover pool's aggregate loan amount is at least as high as the nominal amount of outstanding covered bonds. A failure to meet this requirement triggers cover pool liquidation and the repayment of outstanding covered bonds. For more details, see Koppmann [102], pp. 294–295.

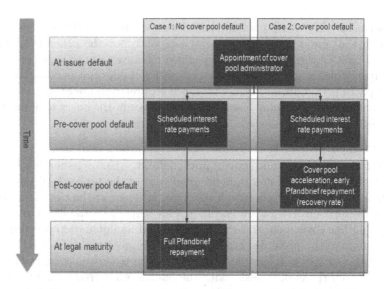

Figure 2.4: Situation post-issuer insolvency (Source: Spangler and Werner [145], p. 19).

2.2 Risks from an Investor's Perspective

From the investor's point of view there are several types of risks arising from a Pfandbrief investment. In the first place, the Pfandbrief holder is exposed to all kinds of risk related to the issuer (*issuer risks*) as the latter is responsible for making full and timely Pfandbrief payments. Upon issuer insolvency, the situation changes and cover pool performance becomes important as payments to the Pfandbrief holder are made out of the cover pool (*cover pool risks*). There are also risks associated with the *timing of Pfandbrief repayments, structural and legal risks* in the context of issuer and cover pool insolvency and *other risks* that do not fit into the aforementioned categories. While, for obvious reasons, issuer risks only matter pre-issuer insolvency, the Pfandbrief holders become exposed to cover pool risks, the risk of timely Pfandbrief repayment and structural and legal risks after issuer insolvency only. Table 2.1 summarizes the relevance of the above discussed risk types prior to and after issuer insolvency. In the following, we shortly discuss the most important of these risks. For a detailed explanation and an in-depth analysis, we refer to Spangler and Werner [145].

Table 2.1: Relevance of Pfandbrief risks prior to and after issuer insolvency (cf. Spangler and Werner [145], p. 30).

Type of risk	Pre-issuer insolvency	Post-issuer insolvency
Issuer risks	✓	-
Cover pool risks	-	✓
Timely Pfandbrief repayment	-	✓
Structural and legal risks	-	✓
Other risks	✓	✓

Issuer risks. The Pfandbrief issuer is responsible to make full and timely Pfandbrief payments. In addition, his activities in the context of cover pool management and the maintenance of cover requirements contribute to the overall quality of the cover pool, which constitutes the investor's secondary source of recourse. Issuer default risk and cover pool management risk therefore play an important role for the Pfandbrief investor. The relevance of *issuer default risk* has been underlined by several Pfandbrief bank bailouts during the last 10 years. It is mainly influenced by the issuer's business model, his creditworthiness and his access to funding sources. While issuer default itself does not necessarily mean that there are losses to the Pfandbrief holder, it triggers the investor's direct exposure to cover pool risk, the risk of timely Pfandbrief repayment and structural and legal risks, which become relevant upon issuer default, see Table 2.1 above. *Cover pool management risk*, on the other hand, is related to uncertainties with respect to the cover pool's future quality and comprises substitution risk and risks in the context of the maintenance of voluntary overcollateralization. *Substitution risk* arises from the fact that, even though the PfandBG defines strict eligibility criteria and cover requirements, the issuer still has some freedom to change the composition of the cover pool and therefore its risk profile. As a consequence, the cover pool's quality cannot be assumed to be constant over time and it is not independent from the issuer's creditworthiness. The closer the bank is to default, the more important substitution risk becomes. The issuer may generally have less high-quality assets or be forced to use good quality cover pool assets for other purposes and replace them by lower quality assets. Furthermore, he might not be willing or able any more to replace bad performing assets in the cover pool which still fulfil the eligibility criteria. In the context of *voluntary overcollateralization*, the main issue is its voluntary nature: there are almost no restrictions for the issuer to suddenly reduce it.

This becomes especially relevant in a situation of financial distress, when the bank might have no other choice but to reduce voluntary overcollateralization to ease funding pressure. Voluntary overcollateralization in excess of legal requirements can therefore not be relied upon. There are also risks associated with the *issuance of further Pfandbriefe* (apart from aspects that are already covered by substitution risk and the maintenance of overcollateralization) and *other issuer-related risks*, which include operational risk, reputation risk and strategic risk. These risks are, however, not Pfandbrief-specific or of secondary importance only.

Cover pool risks.   The quality of the cover pool also plays an important role for the Pfandbrief investor as Pfandbrief payments are made out of the proceeds of the cover pool once the bank has defaulted. The closer the issuer is to insolvency, the more important become the cover pool and the risks associated with it. Even though the PfandBG stipulates strict quality criteria and cover requirements, there are still remaining cover pool risks. The detailed regulations of the PfandBG can mitigate individual (i.e. program-specific) cover pool risks to a considerable extent but, like other covered bond laws, the PfandBG cannot rule out the impact of systemic risks such as a collapse of the (local) financial system or a general deterioration of cover pool asset quality. Under extreme market conditions, cover pool risk can therefore become a big issue. *Refinancing risk* results from asset liability maturity mismatches. Cover pool assets are typically longer dated than outstanding Pfandbriefe, i.e. cash inflows from the natural amortization of cover pool assets may not be sufficient to make scheduled Pfandbrief payments, resulting in a need to raise additional funds. In a stressed market environment such as in the immediate aftermath of an issuer default, it might be difficult to raise funds and asset sales may only be possible at large discounts (*market value risk*), especially when the market is aware of the cover pool's refinancing pressure. Not only interest rate and currency risk (see below) but also *credit spread risk* play an important role in this context. Aspects like asset downgrade risk and changes in credit spreads due to changed risk aversion in the market which are not relevant from a pure cash flows perspective may still have an impact on realisable sale prices. *Interest rate risk* and *currency risk* refer to losses due to unfavourable market movements and arise from interest rate and currency mismatches between the cover pool and outstanding Pfandbriefe. They may lead to insufficient cover pool cash flows or, in a forced-sale-situation, in mark-to-market losses. For German Pfandbriefe, interest rate risk is more important than currency risk as outstanding Pfandbriefe are to a large extent denominated in euros and cover pool assets typically show high concentrations to borrowers in Germany and euro countries. Furthermore, *asset default risk* results from potential losses caused by defaults of cover pool assets. In the case of Mortgage Pfandbriefe, this also contains aspects of *real estate risk* as the recovery value of a defaulted mortgage asset depends on the realizable sale price of the charged property. *Reinvestment risk, prepayment risk* and *counterparty (credit) risk* only play a secondary role for German Pfandbriefe.

Risk of timely Pfandbrief repayment. Investors are also exposed to the risk that the Pfandbrief is repaid earlier or later than the contractual maturity date. Under normal circumstances, *early redemption* and *Pfandbrief extension* can only occur when both the issuer and the cover pool are insolvent. This is, however, already the worst case scenario, where losses caused by issuer and cover pool risks are likely to be more material than losses resulting from lower earned or higher paid interest rates. These risks are therefore not explicitly considered in the following.

Structural and legal risks. Structural and legal risks refer to uncertainties in the context of cover pool segregation, the transition of tasks to the cover pool administrator and the decision of cover pool liquidation. They are intensified by the fact that until today the working of the Pfandbrief mechanics has not yet been tested in practice. Even though the provisions of the PfandBG provide a high level of comfort, there is still some remaining uncertainty as the sole existence of a legal framework does not necessarily mean that bankruptcy remoteness and cover pool continuation are also feasible. The *protection of voluntary overcollateralization* in the context of segregation risk and the *timing of cover pool insolvency* are especially relevant in this context, as they can have a considerable impact on the Pfandbrief's recovery value. The cover pool itself is insolvency-remote by law, but there is still some remaining uncertainty with respect to the protection of voluntary overcollateralization, caused by a lack of explicit legal provisions with respect to its bankruptcy-remoteness. If voluntary overcollateralization (or parts of it) were released to unsecured creditors before all outstanding Pfandbriefe had been repaid, Pfandbrief holders could lose credit enhancement otherwise available to them. In addition, the Pfandbrief holder is exposed to a considerable amount of uncertainty in the context of cover pool liquidation. This is due to the fact that the

PfandBG does not define specific criteria or trigger events regarding the declaration of cover pool insolvency, which means that there is some freedom regarding the decision when cover pool insolvency is to be declared (cover pool insolvency timing). Furthermore, there are no restrictions regarding the amount of assets that can be liquidated at once for the repayment of a specific Pfandbrief maturity. This results in repayment risk from *time subordination* for later maturing Pfandbriefe. The gap between an asset's market value and its intrinsic value is an important aspect to be considered in this context. The decision whether to sell or retain a specific asset is not always straightforward. While asset liquidation brings in cash and reduces cover pool risks and operational complexities, it might lead to significant discounts to the intrinsic value (market value risk) and eliminates any upside potential due to favourable future market conditions. Asset retention, on the other hand, does neither eliminate upside nor downside risk, i.e. the final result may be superior or inferior to immediate liquidation, depending on future market conditions. *Segregation risks* (apart from the protection of voluntary overcollateralization) and risks in the context of *transition to alternative management* are of secondary importance only.

Other risks.   There are also other Pfandbrief-related risks the investor is exposed to, which include *country risk, operational risk*[4], *settlement and related risks, reputational risk* and *regulatory risk*. With the exception of *sovereign risk* which is part of country risk, these risks are not specific to the Pfandbrief or of secondary importance only. In the case of German Pfandbriefe, sovereign risk can currently be neglected.

Figure 2.5 summarizes the main Pfandbrief risks from an investor's perspective as derived based on the findings in Spangler and Werner [145]. It becomes apparent that bank solvency and cover pool performance as well as well as the issuer's activities in the context of cover pool management play an important role. Certain structural and legal risks also need to be considered.

---

[4]Note: Operational risk from Pfandbrief-related business activities (e.g. the cover pool administrator's activities) does not include issuer-related operational risk, which we consider to be part of *other issuer-related risks.*

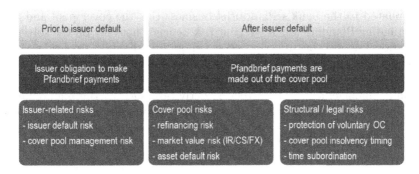

Figure 2.5: Main Pfandbrief risks from an investor's perspective.

2.3 Implications for Modelling

A realistic Pfandbrief model should account for the most important product characteristics and be able to adequately reflect its main risks. In Section 2.1 we found the Pfandbrief's main features to be its dual protection mechanism with full recourse to the issuer and an additional priority claim on the high quality cover pool, the mandatory overcollateralization arising from the legal provisions, the active cover pool management under the issuer, the ring-fencing and continuation of the cover pool upon issuer insolvency and the distinction between the events of issuer and cover pool default. From Section 2.2 we further know that the relevance of risks depends on the occurrence of the event of bank default and that issuer solvency and cover pool performance as well as the issuer's activities in the context of cover pool management are important in the context of risks. In addition, structural and legal risks play a role.

For realistic Pfandbrief modelling we need to distinguish between the events of issuer and cover pool default and consider the timing of these events. Given the important role of the market environment, the model should be able to relate the default events to the economy. More specifically, the impact of stochastic risk drivers (e.g. interest rates and asset creditworthiness) on asset present values and cash flows has to be modelled. Consequently, a simplified version of the bank's balance sheet is needed which accounts for the bank's risk profile (asset riskiness and sensitivity with respect to risk drivers, different asset and liability maturities and resulting asset liability mismatches) and to allow for an explicit distinction of the balance sheet positions which are related to the bank's Pfandbrief business (i.e. the cover pool and the outstanding Pfandbriefe). As default risk depends not only on the solvency situation but also on funding access, illiquidity should be considered as a second default reason in addition to overindebtedness. Furthermore, the dynamic nature of the cover pool requires the specification of asset liability management (ALM) related activities, such as the maintenance of mandatory overcollateralization and potential funding and reinvestment activities. Considering asset present values only is not sufficient for this purpose. The prevalence of cover requirements and funding activities requires the modelling of the bank's cash flow profile. Pfandbrief payments under different scenarios (i.e. pre and post-issuer insolvency, pre- and at cover pool default) also need to be distinguished, taking into account the ring-fencing and cover pool continuation upon issuer default and the Pfandbrief acceleration in case of cover pool insolvency. Finally, the priority of paymens in case of default, including the Pfandbrief holder's priority claim on the cover pool and his residual claim against the issuer's general insolvency estate, have to be adequately reflected. A one-period model is obviously not able to capture all these aspects; a multi-period time setting is needed. Table 2.2 summarizes our required modelling features. In the next chapter we will review the existing credit risk literature with particular focus on these requirements.

Table 2.2: Required features for Pfandbrief modelling.

Model component	Feature
General	Multi-period time setting
Default modelling	Distinction of issuer and cover pool default
	Distinction of overindebtedness and illiquidity
	Link between default and market environment
Balance sheet	Adequate representation of risk profile
	Distinction of Pfandbrief business
	Modelling of asset present values *and* cash flows
Market environment	Stochastic risk drivers
	Impact on asset present values and cash flows
ALM	Active cover pool management
	Funding and reinvestment activities
Pfandbrief mechanics	Dual protection mechanism
	Maintenance of legal cover requirements
	Ring-fencing upon issuer default
	Priority of payments in case of default

3 Credit Risk Models: A Literature Review

This chapter is dedicated to the modelling of default events and gives an overview of the existing credit risk literature with particular focus on bank defaults and Pfandbrief modelling requirements. In Section 3.1 we introduce the two main approaches to credit risk modelling, the structural and the reduced form approach, and describe their areas of application. As the structural approach turns out to be more suitable for our purpose, we have a closer look at structural credit risk models in Section 3.2. Starting from the seminal work of Merton, we explain the basic idea behind the approach and discuss several model specifications and extensions which can be found in the literature. We also investigate the empirical performance of structural credit risk models in the context of credit spread and default risk estimation. In the structural model setup, default is often interpreted as overindebtedness, while liquidity does not play a role. Funding access and liquidity management are, however, very important for banks. In Section 3.3 we therefore deal with the application of structural credit risk models to banks and analyse the implications of different funding assumptions. We also present examples of structural credit risk models which are especially designed for banks. The two most promising ones in the context of Pfandbrief modelling are discussed in more detail. Section 3.4 summarizes the main findings of our literature review. It turns out that none of the discussed models fulfils all our requirements for Pfandbrief modelling.

For a general introduction to credit risk models, we refer to standard textbooks such as Bluhm et al [21], Arvanitis and Gregory [11], Duffie and Singleton [52], Schönbucher [142], Lando [108] or Bielecki and Rutkowski [18]. A comprehensive survey on structural and reduced-form models can be found in Uhrig-Homburg [151], and a detailed overview of structural credit risk models is given in Elizalde [56] and Laajimi [104]. Elizalde [55] covers reduced-form models.

© Springer Fachmedien Wiesbaden GmbH, part of Springer Nature 2018
M. Spangler, *Modelling German Covered Bonds*, Mathematische Optimierung und Wirtschaftsmathematik | Mathematical Optimization and Economathematics, https://doi.org/10.1007/978-3-658-23915-2_3

3.1 Approaches to Credit Risk Modelling

The literature on credit risk modelling and default-claim pricing distinguishes two main approaches: structural models and reduced-form models. *Structural models* link the economic fundamentals of a company (i.e. its capital structure and the value of its assets) and its creditworthiness in an explicit manner. An exogenous firm value process is specified and default occurs once this process hits a given barrier. In structural credit risk models, default probabilities are determined endogenously through the evolution of the firm value process, while recovery rates are either given exogenously or determined endogenously. Prominent examples of structural credit risk models are Black and Cox [19], Longstaff and Schwartz [116], Kim et al [100], Briys and De Varenne [25], Ho and Singer [85], Leland [111], Leland and Toft [113] and Geske [76].

Reduced-form models, on the other hand, rely on an exogenous intensity process to specify the default event. Default is typically triggered by the first jump of this process. The parameters governing the process are calibrated to market data (e.g. credit spreads), but the economics behind the process are unknown. As a consequence, there is no explicit relationship between the default event and the company's economic conditions and default can occur for arbitrary asset values, i.e. it is not predictable and comes as a surprise. Examples of these kind of models are Jarrow and Turnbull [98], Jarrow et al [96] and Duffie and Singleton [51]. Reduced-form models can be interpreted to be a special case of structural models under incomplete information. For more details see Elizalde [57].

The main difference between the structural and the reduced-form approach is the way the default event is specified. For pricing purposes, especially in the case of complex credit derivatives, reduced form models are often the preferred choice as they are more flexible and easier to calibrate. For more fundamental analyses, e.g. a quantification of the impact of economic risk drivers on the company's creditworthiness and resulting consequences for debt holders of different seniority, some kind of structural model is needed. Table 3.1 summarizes the main features of structural and reduced-form credit risk models. Given our requirements for Pfandbrief modelling as derived in Chapter 2, the structural approach is the natural choice. In the following, we will therefore focus on structural credit risk models.

Table 3.1: Features of structural and reduced-form credit risk models.

Feature	Structural models	Reduced-form models
Default specification	Exogenous firm value process hits a given boundary	First jump of an exogenous intensity process
Default vs. economy	Close link	No explicit link
Preferred application	Fundamental analyses	Pricing (complex products)

3.2 Introduction to Structural Credit Risk Models

Structural credit risk models are based on the seminal work of Merton [124]. Most of these models assume that default occurs once a given state variable representing the company's economic or financial conditions falls below a certain boundary. Structural credit risk models differ, among others, with respect to the modelling of this state variable and the choice of the default barrier. In Sections 3.2.1 and 3.2.2 we explain the basic idea behind Merton's model and discuss selected model extensions which can be found in the literature. In Section 3.2.3 we look into the specification of the default barrier, and Section 3.2.4 focuses on the role of liquidity in structural credit risk models. The performance of structural credit risk models in the context of credit spread and default estimation is discussed in Section 3.2.5.

3.2.1 The Merton Model

In his seminal work, Merton [124] assumes perfect, competitive and frictionless markets[1], continuous trading of assets and a constant risk-free interest rate r. Under the real-world measure, the company's asset value V follows a lognormal diffusion process

$$dV(t) = (\mu V(t) - \delta)\, dt + \sigma V(t) dW(t),$$

with μ being the instantaneous expected asset return per unit of time, δ the firm's total dollar payout per unit of time, σ the constant volatility of the firm value return and $(W(t))_{t \geq 0}$ a standard Brownian motion.

---

[1]This means that there are no transaction costs or taxes, assets are liquid, borrowing and lending is always possible at the same rate, there are no problems with indivisibilities of assets and restrictions with respect to short selling do not exist.

Merton considers a company with a very simplistic debt structure, with two classes of claims only: a single homogeneous class of debt, represented by a risky zero coupon bond with face value F and maturity T, and equity (the residual claim). The debt structure in his model is fixed, meaning that no further debt can be issued. Merton also assumes that default can only occur at the zero coupon bond's maturity T. Whether or not the company defaults at this time depends on the then-prevailing asset value $V(T)$. If it is sufficient to repay the outstanding debt, $V(T) \geq F$, the company remains solvent, the full face value F is repaid to bondholders and equity holders receive a payment of $V(T) - F \geq 0$. If the value of the assets falls below F, the company defaults. In this case, an amount of $V(T)$ is repaid to bondholders and they experience a loss of $F - V(T) > 0$, while equity holders receive nothing. In Merton's model, the time-T payoffs to debt (D) and equity (E) are therefore given by

$$D(T,T) = \min\left(F; V(T)\right) = F - \max\left(0; F - V(T)\right),$$
$$E(T) = \max\left(0; V(T) - F\right),$$

with $D(T,T) + E(T) = V(T)$. The time-T payoffs $D(T,T)$ and $E(T)$ can be interpreted as contingent claims on the firm's asset value $V(T)$. Holding the risky zero coupon bond corresponds to having entered a long position of an otherwise identical but default-free zero coupon bond with face value F and maturity T, and a short position of a European put option on V with expiry T and strike F. The equity holders' position, on the other hand, resembles a call option on the firm's asset value with expiry T and strike F. The time-t prices of debt and equity are then obtained by calculating the risk-neutral expectation of the discounted final payoffs,

$$D(t,T) = P(t,T) \cdot F - P(t,T) \cdot \mathbb{E}^{\mathbb{Q}}\left[\max\left(0; F - V(T)\right) \,\Big|\, \mathcal{F}_t\right],$$
$$E(t) = P(t,T) \cdot \mathbb{E}^{\mathbb{Q}}\left[\max\left(0; V(T) - F\right) \,\Big|\, \mathcal{F}_t\right],$$

(3.1)

with $P(t,T) := e^{-r \cdot (T-t)}$ and $\mathbb{E}^{\mathbb{Q}}\left[\bullet | \mathcal{F}_t\right]$ being the conditional expectation under the risk-neutral measure $\mathbb{Q}$ given the information at time t. Using results from option pricing theory, cf. Black and Scholes [20], the calculation of the risk-neutral expectations in (3.1) is straightforward.

3.2.2 Extensions of the Merton Model

As Merton's model relies on quite restrictive assumptions, it has been extended in many ways. These extensions concern, amongst others, Merton's assumptions regarding the company's debt structure, the timing of default, the firm's asset value process, the modelling of interest rates and recovery rates and the costs of default, see Figure 3.1.

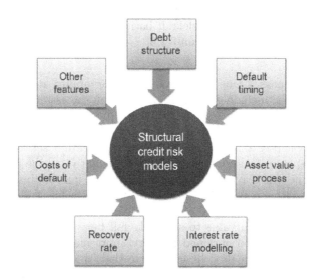

Figure 3.1: Selected extensions of Merton's model.

Debt structure. Merton assumes a very simplistic debt structure with only one class of debt, represented by a risky zero coupon bond of finite maturity. In practice debt structures are much more complex and payoffs in case of default may be specified by contractual provisions. Black and Cox [19] handle subordination within the option pricing framework by adjusting payment formulas accordingly, but they only allow for one common maturity date. Geske [76] accounts for a more complex capital structure: his model works for multiple debt issues with different maturities and coupons. In the context of Pfandbrief modelling, flexibility with respect to the debt structure is needed, to allow for an adequate representation of the Pfandbrief bank's risk profile and a distinction between Pfandbriefe and other liabilities.

Default timing. In Merton's model, default can only occur at a fixed debt maturity which is not realistic. So-called *first-passage-time models*, pioneered by Black and Cox [19], have relaxed this assumption. These models assume that default occurs once a given state variable representing the company's economic and financial conditions, typically represented by the value of the firm's assets, falls below a given default barrier. The choice of this barrier is discussed in more detail in Section 3.2.3.

Asset value process. Due to its analytical tractability, many structural credit risk models, including Merton [124], rely on the assumption that the firm's asset value follows a lognormal diffusion process with constant volatility. This implies that credit spreads for short maturities converge to zero (cf. Lando [108], Chapter 2.2.2), which contradicts empirical observations of positive market spreads for short-term debt and has motivated the introduction of jumps in the firm's asset value process. Under a jump diffusion process, default occurs either expectedly (if the default barrier is hit as a consequence of steady declines during times of normal fluctuations) or as a surprise (if a jump causes the firm value to suddenly fall below the critical barrier). Positive spreads can be obtained for very short maturities and even for firms with high credit quality. In addition to that, the endogenously created recovery rates are random and depend on the firm's asset value at the time of default. A disadvantage of the introduction of jumps is that it comes at the expense of analytical tractability and complicates calculations considerably. Examples of structural credit risk models with jumps in the asset value process are Schönbucher [141], Zhou [169], and Mason and Bhattacharya [117]. In the context of Pfandbrief modelling, flexibility with respect to the choice of the asset value process is needed, to account for the fact that the assets on a bank's balance sheet typically consist to a large part of risky bonds and loans for which a lognormal diffusion process is not suitable (see discussions in Sünderhauf [148], p. 81).

Interest rates. Several models such as Kim et al [100], Longstaff and Schwartz [116], Briys and de Varenne [25] and Hsu et al [87] relax Merton's constant interest rate assumption by introducing stochastic interest rates. Such approaches are more realistic and allow for the modelling of dependencies between the company's asset value and the risk-free interest rate. As in the case of jumps in the firm's asset value, the introduction of stochastic interest rates reduces analytical tractability and complicates modelling. Nevertheless, stochastic interest rates are necessary for realistic Pfandbrief modelling.

Recovery rates. Merton's model implicitly assumes that the recovery payment, which equals $V(T) < F$, is determined endogenously and depends on the company's remaining asset value at time T. A similar assumption is also made by Black and Cox [19], Geske [76] and Vasicek [155]. In their models, the recovery payment equals $V(\tau)$, i.e. the remaining asset value at the default time τ. Other models, such as Kim et al [100] and Longstaff and Schwartz [116], assume an exogenously given recovery rate (typically a percentage fraction of outstanding debt) which implies that the payment in case of default is independent from the then-prevailing asset value. As opposed to exogenous recovery rates, endogenous recovery rates allow for an explicit modelling of dependencies between default probabilities and recovery rates.

Costs of default. Merton assumes that the full asset value is available to repay bondholders in case of default. This assumption is not realistic due to bankruptcy and asset liquidation costs which occur in practice.[2] Such costs can be incorporated into structural credit risk models by assuming that only a fraction $\alpha \in [0, 1]$ of the asset value is left upon default, while a fraction $(1 - \alpha)$ is lost. Models accounting for costs of default are, for example, Leland [111], Uhrig-Homburg [150] and François and Morellec [72].

Further model extensions. In the literature, there are many other extensions of the Merton [124] model. Among them one finds models with dynamic capital structures (as in Collin-Dufresne and Goldstein [30], Fischer et al [71], Goldstein et al [79] and Ju et al [99]), models accounting for strategic default and out-of-court renegotiations (see Anderson and Sundaresan [9], Anderson et al [10], Fan and Sundaresan [70], Mella-Barral and Perraudin

---

[2]Forced asset liquidation may result in high liquidation costs, depending on the market environment and asset liquidity. In this context, the so-called liquidity premium in credit spreads plays an important role. In times of crises when liquidity in financial markets decreases, financial institution may also be confronted with unexpectedly high bid-ask spreads when forced to sell or liquidate positions.

[123] and Mella-Barral [122]), and liquidation process models[3] (e.g. Moraux [126], Galai et al [74], Paseka [134] and Broadie et al [26]). In the following, we do not further consider these kind of model extensions as they are of low relevance for Pfandbrief modelling.

3.2.3 Specification of the Default Barrier

The majority of structural credit risk models assume that default occurs once the firm's asset value falls below a given barrier. In the credit risk literature, two different types of default barriers are distinguished: endogenous and exogenous ones. Models with *endogenous default barriers* as for example Leland [111], Leland and Toft [113] and Geske [76] assume that equity holders choose the timing of default such that the value of their claim is maximized. At each point in time they decide whether or not it is worth to make scheduled debt payments by comparing the benefits of keeping the company alive to the costs of making these payments. By doing so, they deduct their default barrier endogenously and stop making payments once the asset value falls below this threshold. This implies that debt service payments may continue to be made even though the value of the company's assets is below the principal of debt or there are not sufficient cash inflows. Endogenous default models look at certain aspects of the bankruptcy process in more detail and allow for what Laajimi [104] calls "a richer modeling of the default decision". Depending on the model setup it might, however, be quite complex to derive the endogenous default barrier as an optimization problem needs to be solved at each point in time. The more complicated the asset value process and the debt structure are chosen, the more difficult this task becomes. As Pfandbrief modelling requires non-standard assumptions regarding the bank's debt structure and asset value process, we do not consider endogenous default barriers in the following. Instead, we focus on the more intuitive exogenous default barriers.

As suggested by their name, *exogenous default barriers* are given exogenously, and they are typically somehow related to the face value of debt. The barrier itself can be constant as in Longstaff and Schwartz [116] or Kim et al [100], time-dependent as in Black and Cox [19], or stochastic as in Briys and de Varenne [25] and Hsu et al [87]. It is often interpreted as a safety covenant that protects bondholders from losses when the firm's asset value

---

[3]Liquidation process models account for the fact that default results not necessarily in immediate liquidation but can be the outcome of a long lasting process.

decreases too much. As opposed to endogenous default barriers, exogenous default barriers are comparably easy to determine and more in line with the interpretation of *overindebtedness*. In the following, we consider a simplified debt structure with only one class of debt, represented by a risky zero coupon bond with face value F and maturity T. The most simplistic choice is to set the default barrier equal to the face value of debt as in the Merton model, $V_B^C := F$. A more sophisticated alternative is the time-dependent, potentially stochastic, version which can be found in Black and Cox [19] and Briys and de Varenne [25]:

$$V_B^\alpha(t) := \alpha \cdot F \cdot P(t, T), \quad 0 \le \alpha \le 1. \tag{3.2}$$

Here, $P(t, T)$ is the time-t value of a default-free zero coupon bond maturing in T, and α is an exogenously given constant fraction. The closer α is to one, the more protective this barrier is. The default barrier $V_B^\alpha(t)$ decreases for increasing maturities T as $P(t, T)$ becomes smaller. In case of stochastic interest rates, $P(t, T)$ also becomes stochastic, meaning that it is not known at times $s < t$. One problem associated with this default barrier is that for $\alpha < 1$ it can happen that $V_B^\alpha(T) < V(T) < F$, meaning that the asset value at maturity is above the barrier $V_B^\alpha(T) = \alpha \cdot F$ and below the face value of debt. This implies that the company does not default, but the value of its assets is not sufficient to repay the maturing debt. To avoid unwanted implications, liability payoffs have to be defined such that this is taken into consideration, see Briys and de Varenne [25]. Another alternative is the default barrier used by Moody's KMV:[4]

$$V_B^{KMV}(t) := F_{STD}(t) + 0.5 \cdot F_{LTD}(t), \tag{3.3}$$

with $F_{STD}(t)$ being the face value of the company's outstanding short-term debt and $F_{LT}(t)$ the face value of its outstanding long-term debt, both as seen from time t. This barrier accounts for the fact that short-term debt has to be repaid soon, while long-term debt does not require nominal repayments to be made in the near future. According to Leland [112], KMV considers "short-term" to be the time horizon for which the default probability is calculated. The longer this time horizon, the higher the fraction of short-term debt and therefore the default barrier.

---

[4]For more details on the KMV model, see Crosbie and Bohn [35].

The specification of the default barrier is not straightforward[5] and can have a large impact on model results. Empirical evidence suggests that the actual asset value at which the firm defaults is lower than the face value of debt and lies somewhere in between short-term debt and total liabilities, cf. Crosbie and Bohn [35]. This supports the choice of a default barrier as defined in Equation (3.3). Davydenko [36], who uses market values of defaulting companies, finds mean (median) default barrier estimates of 66% (62%) of the face value of outstanding debt. This is roughly in line with Huang and Huang [88], who assume a default boundary of around 60%. Results obtained by Fabozzi et al [69] suggest that the timing of debt payments should also be accounted for when specifying an exogenous default boundary. They compare six different structural credit risk models including Merton [124], Black and Cox [19], Longstaff and Schwartz [116] and a flat barrier model and find that an exponentially decaying barrier as in Black and Cox [19] clearly dominates a flat barrier. According to them, the Black and Cox model is in general the most dominant model out of the six ones considered, which suggests a default barrier as in Equation (3.2). For Pfandbrief modelling we will rely on a modified mixture of the default barriers in (3.2) and (3.3). For more details we refer to Section 4.5 below.

3.2.4 The Role of Liquidity

As structural credit risk models assume that default occurs once the firm's asset value falls below a given threshold, default can be interpreted as some kind of *overindebtedness*. However, empirical findings indicate that in practice overindebtedness is not the only default reason and that financing frictions also play an important role in the context of default. For more details, see Davydenko [36] and Dionne and Laajimi [49]. Realistic modelling should, therefore, not only account for overindebtedness but also for illiquidity.

Structural credit risk models mostly ignore the role of liquidity. Notable exceptions are so-called *cash based models* such as Kim et al [100], Anderson and Sundaresan [9] and Anderson et al [10]. These models assume that default occurs once an exogenously given cash flow falls below current debt service payments, which implies that default may occur even though the

---

[5] As pointed out by Dionne and Laajimi [49], neither the dynamics nor the location of the default barrier are visible, meaning that it has to be specified based on indirect information.

company is not overindebted. This way of modelling cash shortages is not very convincing as it completely rules out the possibility to raise new funding, i.e. the costs of new issuances are assumed to be infinitely high. In addition to that, the resulting analytics are often equivalent to value-based models when the firm's asset value is modelled as a fixed multiple of the firm's cash flow, cf. Leland [112].

A second type of structural models which implictely accounts for liquidity are *endogenous models with frictionless markets*, including those by Leland [111] and Leland and Toft [113], which do not distinguish between cash shortage and overindebtedness. They assume that there are no market frictions, meaning that the company can always raise additional equity to overcome liquidity problems, at least as long as enough value remains to motivate the issuance of new equity. Such costless equity issuance implies that liquidity has no value, which is not in line with empirical evidence, cf. Asvanunt et al [12]. The truth probably lies somewhere in between the two extreme modelling assumptions of cash based models and models with frictionless markets. Cash shortages do not always lead to an immediate default nor can they always be overcome with certainty.

Only few structural models incorporate both overindebtedness and illiquidity as default reason. This is due to the fact that modelling both types of default events simultaneously is challenging and results in increased model complexity (Davydenko [36]). Two models which account for both default events are the ones by Liang et al [114] and Uhrig-Homburg [150], which will be discussed in more detail in Section 3.3. As banks typically heavily rely on external funding and liquidity management, modelling default due to illiquidity is very important in the context of Pfandbrief modelling.

3.2.5 Empirical Findings on Model Performance

Various empirical studies focus on the ability of structural credit risk models to predict credit spreads and default probabilities. Although most of these works exclude financial institutions, see Imerman [93], p. 9, they still provide valuable insights into model behaviour and the quality of predictions. In the following, we give a short overview of the main findings in this context. For a more comprehensive review of empirical findings with respect to credit spread prediction and an in-depth discussion of the suitability of structural credit risk models for default risk assessment, see Sünderhauf [148], pp. 67–77.

One common finding with respect to *credit spread prediction* is that structural models do not perform well and tend to underestimate observed credit spreads, especially for short maturities. This poor performance is often explained by empirical evidence that default risk is not the sole driver of credit spreads. Eom et al [59] observe that credit spreads from the Merton model are too low as compared to empirical spreads and the studies by Elton et al [58], Huang and Huang [88], Ericsson and Reneby [60], Perraudin and Taylor [135] and Collin-Dufresne et al [31] suggest that a large proportion of credit spreads is driven by other factors than default risk, namely liquidity, risk premia and tax effects. Based on these findings, Sünderhauf [148] argues that structural credit risk models cannot be expected to explain the full credit spread observed in the market but only the default risk part of it.

In the context of *default prediction*, structural credit risk models seem to perform better.[6] Sünderhauf [148] refers to several studies which provide evidence that the default risk itself is acceptably explained by structural credit risk models. Fabozzi et al [69], who study six structural credit risk models (Merton [124], Black and Cox [19], Leland and Toft [113], Longstaff and Schwartz [116], Geske [76] and a flat barrier model), find that these models are quite robust against sample selection and conclude that structural models are suitable for default prediction. Leland [112], who compares the Longstaff and Schwartz [116] and the Leland and Toft [113] model, finds that long-term default probabilities are predicted quite accurately, while short-term default probabilities tend to be underestimated. He draws the conclusion that a jump component should be included in the asset value process. All in all, the structural approach to credit risk modelling seems appropriate for our purpose, as our focus is on default modelling and not on the replication of market spreads or prices.

3.3 Bank Default Modelling

Even though there is a vast amount of literature on structural credit risk models, only few of these works deal with their application to banks. As pointed out by Chen et al [28], many structural credit risk models cannot capture the refinancing strategies such as short-term debt roll-over, the complex liability structure and the high leverage which are typical for banks.

---

[6] As opposed to credit spreads, default probabilities are not considered to be impacted by additional factors such as liquidity, risk premia and tax effects (Leland [112]).

Nevertheless, in the context of Pfandbrief modelling we have to deal with the application of structural credit risk models to banks. In Section 3.3.1 we therefore analyse potential bank funding and liquidity management strategies. In Section 3.3.2 we then present examples of existing structural models which are particularly designed for banks. The two most promising ones, Sünderhauf [148] and Liang et al [114], are discussed in more detail.

3.3.1 Funding and Liquidity Management

In practice, bank survival strongly depends on the bank's refinancing strategy and its ability to obtain funding when needed, which is not just a matter of costs. Depending on the market environment and the issuer's specific situation, access to funding sources might be restricted or not possible at all. This is why banks try to reduce their liquidity risk exposure by diversifying funding sources. In the following, we discuss several bank funding and liquidity management options as shown in Figure 3.2, focusing on their relevance in practice, their presence in structural credit risk models and potential modelling implications.[7]

Unsecured funding. Liquidity needs can be met by engaging in unsecured funding activities such as overnight borrowings and the issuance of certificates of deposit or commercial papers. Under normal conditions, the availability of unsecured funding may be a viable assumption, but in case of a bank-specific acute funding need or a market-wide funding crisis, unsecured funding cannot be relied upon any more. As pointed out by Matz [119], "prudent bankers assume that unsecured funding sources are only available under almost benign conditions". Structural credit risk models which assume that the debt structure is static cannot account for the issuance of new debt at all. The model by Leland [111] and its extensions incorporate the issuance of new debt to fund contractual payments by assuming that debt is continuously rolled over and replaced by new debt, meaning that the company's leverage remains constant over time. This implies that rollover risk is completely ignored as there are no market frictions and unsecured funding is always available.

---

[7]We do not explicitly consider securitization as the discussion would be similar to secured funding (in case of synthetic deals) or asset sales (in case of true sale deals).

Figure 3.2: Bank funding and liquidity management options.

Secured funding. Secured funding, which includes repurchase agreements and collateralized central bank open market operations, requires assets to be posted as collateral in exchange for cash. These assets typically have to fulfil certain eligibility criteria and are pledged at a haircut to their present value, reflecting the uncertainty regarding the future realisable sale price due to price volatility and potential liquidation discounts. As pointed out by Matz and Neu [120], secured funding has several advantages. First of all, due to the collateralization it is considered to be more reliable than unsecured borrowings, meaning that lenders can be expected to be less likely to withdraw their funds in case the bank gets into troubles. In addition to that, secured funding may be the preferred alternative to asset sales when realisable prices or accounting-related reasons make it less desirable to sell certain assets. Finally, even less liquid assets can potentially be used to provide liquidity. As long as there are sufficient unencumbered assets available which can be posted as collateral, secured funding can therefore be considered to be a comparably reliable funding option. Still, it cannot be assumed to be available without limitation, see Matz and Neu [120]: "There have been situations where funds providers were unwilling to provide repo funding to banks experiencing funding crisis, even though the banks were willing to pledge high-quality securities as collateral." According to the Basel Committee on Banking Supervision [14], even secured funding with overnight maturity should not be assumed to be rolled over automatically. Structural credit risk models typically do not incorporate secured funding options. One notable exception is the bank-run model by Liang et al [114], which will be discussed in more detail in Section 3.3.2.2 below.

Committed liquidity lines. Committed liquidity lines from other banks can be a flexible way to deal with short-term funding needs. Again, the main problem with this liquidity source is that it cannot be relied upon in a bank-specific crisis, even in the absence of so-called 'material adverse change clauses'[8], as the commitment might be withdrawn.[9] The Basel III calculation rules for the liquidity coverage ratio require that committed credit or liquidity facilities are not considered in a stress scenario (Basel Committee on Banking Supervision [13], p. 36). Committed liquidity lines are rarely found in the credit risk literature. Two notable exceptions are Asvanunt et al [12] and Escobar et al [61]. Both of them do not require the liquidity line to be backed by assets.

Equity issuance. Another funding option is the issuance of additional equity capital. According to Matz and Neu [120] this does, however, not seem to be a notable source of liquidity for banks. Imerman [93], on the contrary, argues that the issuance of additional capital becomes more relevant in times of financial distress when financial institutions have restricted access to debt markets, decide to deleverage or are forced to raise capital by regulators. Structural credit risk models (e.g. Geske [76]) often assume that outstanding debt is repaid by raising new equity capital, which implies continuous deleveraging over time. As banks often rely on debt rollover, this may lead to an underestimation of default risk.

Liquidity reserves. Liquidity reserves are considered to be a very reliable source of cash and can provide a certain buffer for times of funding distress.

---

[8]Material adverse change clauses allow for a withdrawal of the commitment when the borrower's financial condition deteriorates.

[9]See, for example, Matz [119]: "In some previous situations, capital markets' counterparties have even reneged on contractual obligations in the apparent belief that the risk of being sued by the bank if it survives is less than the risk of losing their money".

Basel III requires banks to hold sufficient unencumbered[10] high quality liquid assets, i.e. cash or assets that can be readily converted into cash, to survive at least 30 days in a liquidity stress scenario. Excess cash is typically not hold, cf. Matz and Neu [120]: "Well-managed banks rarely hold much more cash than they require for ongoing operations. Therefore, for a bank that continues in business, cash is a fixed asset." Due to the cost of holding liquidity, liquidity reserves are not the most desirable option. Some structural credit risk models such as Asvanunt et al [12] and Anderson and Carverhill [8] account for liquidity reserves, but they typically consider cash holdings only. In structural credit risk models with perfect frictionless markets, cash holdings do not make much sense as new funding can be raised without costs at any time. The incentive to hold cash only arises from the prevalence of financing constraints.

Asset sales. Another funding alternative available to banks is to sell assets. However, illiquid assets such as loans might be hard to sell, even under normal market conditions. If assets, particularly less liquid ones, need to be sold on short-notice this may not be possible at, or at least close to, their fair value, especially in times of market distress. During a flight to quality it might even not be possible to sell certain previously liquid assets at all. Except for assets belonging to the bank's liquidity reserves, asset sales may therefore not be the most realistic funding option. Structural credit risk models typically do not allow for asset sales.[11] One reason for this may be the increased mathematical complexity under this option, see Lando [108], p. 29. Allowing for asset sales also has implications on the modelled term structure of default probabilities and leads to a redistribution of wealth. A higher priority is assigned to short-term debt at the cost of long-term debt and the value of equity is increased at the cost of debt holders, see Ho and Singer [85] and Morellec [127].

The above discussions reveal that not all assumptions made in the context of funding availability are realistic and that some of them have unwanted implications for modelling. Most structural credit risk models which incorporate funding options either assume the rollover of unsecured debt or the issuance

---

[10]Here, unencumbered means "free of legal, regulatory, contractual or other restrictions on the ability of the bank to liquidate, sell, transfer, or assign the asset" (Bank of International Settlement [13], p. 9).

[11]As an example, take Kim et al [100] who assume that there are bond indentures that prevent the sale of assets.

of additional equity. While the availability of unsecured funding cannot be relied upon, assuming the issuance of new equity leads to an artificial drop of leverage and, as banks typically rely on debt rollover, in an underestimation of default risk. All in all, we consider a collateralized bank liquidity line to be the most adequate assumption for our purpose. This combines the characteristics of secured funding (a reliable funding source) with the ones of a liquidity line (cash can be raised at fixed conditions). To account for the fact that liquidity lines may be withdrawn when conditions deteriorate, we further assume that the provider of the liquidity line has the option to decide whether or not to prolong the granted funding at each point in time. For more details on the explicit setup, see Section 4.4 below.

3.3.2 Structural Credit Risk Models for Bank Default

As already mentioned before, there are only very few structural credit risk models in the literature which are especially designed for banks. In the following, we shortly present four exceptions, the approaches by Imerman [93], Uhrig-Homburg [150], Sünderhauf [148] and Liang et al [114].

Imerman [93] incorporates features from the compound option approach, cf. Geske [76], [75], Geske and Johnson [77], and from Leland-type models as in Leland [111], Leland and Toft [113]. He allows for complex liability structures and permits two different funding options to refinance maturing debt: debt rollover and the issuance of new equity. Imerman does, however, not account for market frictions and limited access to external financing, i.e. he completely ignores potential defaults due to liquidity reasons. In addition to that, the default barrier in his model is chosen endogenously. His approach is therefore not considered further in the following.

Uhrig-Homburg [150] extends the model by Leland [111] and introduces liquidity constraints by assuming non-zero but finite costs of equity issuance for firms close to distress. She explicitly distinguishes between the two default reasons overindebtedness and illiquidity, but debt payments are financed by the issuance of new equity, which implies an artificial deleveraging over time. As the default barrier in her model is also chosen endogenously, her model is not considered further in the following.

The models by Sünderhauf [148] and Liang et al [114], on the other hand, both use exogenous barriers to model default due to overindebtedness and exhibit

further very interesting features in the context of Pfandbrief modelling. Both
models are now discussed in more detail.

3.3.2.1 Sünderhauf's Mortgage Bank Model

The mortgage bank model introduced by Sünderhauf [148] is a one-period
model for Pfandbrief default risk pricing.[12] It considers the specific structure
of a Pfandbrief bank's balance sheet and accounts for the issuer's obligation to
maintain cover requirements and the specific priority of Pfandbrief payments
in case of default. The model is based on standard assumptions in the
context of structural credit risk models such as perfect capital markets, no
arbitrage, rational market participants, the existence of a risk-free investment
opportunity and strict absolute priority.

The bank's balance sheet. Sünderhauf considers a simplified version of
a typical mortgage bank's balance sheet as shown in Figure 3.3. On the
asset side, there is a mortgage cover pool (HDS), a public sector cover
pool (KDS) and a position called *other assets* (SA), while the liability
side comprises Mortgage Pfandbriefe (HPF), Public Pfandbriefe ($ÖPF$), a
position called *other liabilities* (SV) and equity (EK).

Bank balance sheet	
Mortgage cover pool (HDS)	Mortgage Pfandbriefe (HPF)
Public cover pool (KDS)	Public Pfandbriefe (ÖPF)
Other assets (SA)	Other liabilities (SV)
	Equity (EK)

Figure 3.3: Sünderhauf's simplified balance sheet.

---

[12]By *mortgage bank*, Sünderhauf refers to the time prior to the PfandBG when only
specialised mortgage banks were allowed to issue Pfandbriefe under the 'Hypotheken-
bankgesetz', one of the predecessors of the PfandBG.

Sünderhauf assumes that all liabilities mature at the same time T, and that all assets mature at the same time $S \geq T$.[13] In his model, the bank's balance sheet is static until time T, meaning that no further debt is issued and no assets are purchased or sold. He also assumes that no cash flows occur prior to time T and that cover requirements are only reestablished at the liabilities' maturity T. Under these assumptions, the time-t balance sheet equation in terms of present value is given by

$$V^{HDS}(t) + V^{KDS}(t) + V^{SA}(t) = V^{HPF}(t) + V^{\ddot{O}PF}(t) + V^{SV}(t) + V^{EK}(t),$$

for $0 \leq t \leq T \leq S$, with $V^x(t)$, $x \in \{HDS, KDS, SA, HPF, \ddot{O}PF, SV, EK\}$, being the time-$t$ present value of the corresponding balance sheet position.

For the ease of presentation we concentrate on a special case of Sünderhauf's model in the following; the case of only one Pfandbrief type.[14] In this special case, the asset side of the bank's balance sheet simplifies to one cover pool (DS) and one position called other assets (SA), while the liability side consists of Pfandbriefe (PF), other liabilities (SV) and equity (EK). The bank's balance sheet equation at time t then becomes

$$V^{DS}(t) + V^{SA}(t) = V^{PF}(t) + V^{SV}(t) + V^{EK}(t),$$

for $0 \leq t \leq T \leq S$.

Risk drivers. Sünderhauf considers several stochastic risk drivers which have an impact on the value of the bank's assets: the risk-free interest rate, state variables for the creditworthiness of the cover pool and the other assets and the volatilities of these state variable processes. Correlations between the corresponding processes are also taken into account.

- *Risk-free interest rate:* The risk-free short rate under the risk-neutral measure $\mathbb{Q}$ is modelled by an extended Vasicek model as in Hull and White [92]:

$$dr(t) = (\theta_r(t) - \kappa_r r(t))\, dt + \sigma_r dW_r(t), \qquad (3.4)$$

---

[13]He argues that term transformation with liabilities maturing later than assets is of low relevance to mortgage banks, cf. Sünderhauf [148], pp. 81f.

[14]For the setup of the full model, see Sünderhauf [148], Chapter 5.1.

with a standard Brownian motion $(W_r(t))_{t\geq0}$ and positive constants κ_r and σ_r. $\theta_r(t)$ is chosen to exactly fit the interest rate term structure observed in the market,

$$\theta_r(t) := \frac{\partial f^M(0,t)}{\partial t} + \kappa_r f^M(0,t) + \frac{\sigma_r^2}{2\kappa_r}\left(1 - e^{-2\kappa_r t}\right),$$

with $f^M(0,t) := \frac{-\partial \ln P^M(0,t)}{\partial t}$ being the market instantaneous forward rate prevailing at time 0 for the maturity t and $P^M(0,t)$ the time-0 market price of a risk-free zero coupon bond with maturity t. The dynamics in Equation (3.4) imply mean reversion to a long-term, time-depending level $\theta_r(t)$ with speed κ_r. For more details regarding the Hull-White extended Vasicek model, see Brigo and Mercurio [24], Chapter 3.3.

- *State variables for asset creditworthiness:* The risk-neutral dynamics of the state variable processes which drive the bank's asset credit risk are given by jump diffusion processes,

$$\frac{d\tilde{V}^x(t)}{\tilde{V}^x(t)} = (r(t) - \lambda_x\tilde{v}_x)\,dt + \sigma_x(t)dW_x(t) + (J_x(t) - 1)\,d\Pi_x(t), \quad (3.5)$$

for $x \in \{DS, SA\}$, with $(W_x(t))_{t\geq0}$ being a standard Brownian motion under $\mathbb{Q}$ and $(\Pi_x(t))_{t\geq0}$ a Poisson process with intensity λ_x. The jump size of the Poisson process is $J_x(t) > 0$ and $\mathbb{E}^{\mathbb{Q}}[J_x(t)] = \tilde{v}_x + 1$. Jump sizes are assumed to be i.i.d. lognormal with $\ln J_x(t) \sim \mathcal{N}\left(\mu_{J_x}, \sigma_{J_x}^2\right)$, implying $\tilde{v}_x = \mathbb{E}^{\mathbb{Q}}[J_x(t) - 1] = \exp\{\mu_{J_x} + 0.5 \cdot \sigma_{J_x}^2\} - 1$. Furthermore, λ_x, μ_{J_x}, and $\sigma_{J_x}^2$ are constants with $\lambda_x, \sigma_{J_x}^2 \geq 0$, and $\sigma_x(t)$ is stochastic, see specification below. The jump diffusion process in Equation (3.5) is a combination of a geometric Brownian motion and a jump process. As long as no jump occurs, the process evolves according to a geometric Brownian motion. Discrete jumps of random size, drawn from a lognormal distribution, occur with intensity λ_x. To compensate for the influence of these jumps, the drift of the geometric Brownian motion is adjusted by $\lambda_x\tilde{v}_x$.

- *Volatilities of state variable processes:* The volatilities of the state variable processes are assumed to be stochastic. Under the risk-neutral measure $\mathbb{Q}$ they follow the dynamics proposed by Hull and White [91],

$$d\sigma_x^2(t) = \kappa_{\sigma_x}\left(\theta_{\sigma_x} - \sigma_x^2(t)\right)dt + \sigma_{\sigma_x}\sqrt{\sigma_x^2(t)}dW_{\sigma_x}(t), \quad (3.6)$$

for $x \in \{DS, SA\}$, with $(W_{\sigma_x}(t))_{t \geq 0}$ being a standard Brownian motion under $\mathbb{Q}$, and positive constants κ_{σ_x}, θ_{σ_x} and σ_{σ_x}. As in the interest rate case, the dynamics in Equation (3.6) imply mean reversion to a long-term level. Here, the long-term level is given by θ_{σ_x} and the mean reversion speed is κ_{σ_x}.

- *Correlations:* In Sünderhauf's model, the stochastic processes for the state variables and the corresponding stochastic volatility processes are assumed to be correlated according to

$$dW_x(t)dW_{\sigma_x}(t) = \rho_{x,\sigma_x}dt,$$

for $x \in \{DS, SA\}$. The stochastic processes of different state variables as well as the stochastic processes of different volatility processes are also assumed to be correlated, with

$$dW_x(t)dW_y(t) = \rho_{x,y}dt, \qquad dW_{\sigma_x}(t)dW_{\sigma_y}(t) = \rho_{\sigma_x,\sigma_y}dt,$$

for $x, y \in \{DS, SA\}$, $x \neq y$. In addition to that, the risk-free interest rate process is assumed to be correlated to the state variable processes and the processes of the stochastic volatilities,

$$dW_r(t)dW_x(t) = \rho_{r,x}dt, \qquad dW_r(t)dW_{\sigma_x}(t) = \rho_{r,\sigma_x}dt,$$

with $x \in \{DS, SA\}$. The processes Π_x and J_x, on the contrary, are assumed to be uncorrelated to all processes and variables.

For a more detailed discussion of the risk factor modelling assumptions and their implications, see Sünderhauf [148], Chapter 4.1.4.

Asset modelling. Sünderhauf distinguishes three types of asset positions. In the special case of only one Pfandbrief type this reduces to two positions: a cover pool (DS) and other assets (SA). In Sünderhauf's model, the cover pool is modelled as a risky zero coupon bond with nominal N^{DS} and maturity S. Following an idea by Andersen and Cakici [7], he assumes that the cover pool defaults at time S if the value of the state variable associated with the cover pool is below the cover pool's nominal, $\tilde{V}^{DS}(S) < N^{DS}$, and its recovery value is then given by $\tilde{V}^{DS}(S)$. Otherwise, the cover pool survives and the nominal is repaid in full. The time-S cash flow of the cover pool is therefore given by

$$V^{DS}(S) := \min\left(N^{DS}; \tilde{V}^{DS}(S)\right).$$

Sünderhauf argues that modelling the risky cover pool cash flow as in the above equation is more appropriate than just setting $V^{DS}(S) = \tilde{V}^{DS}(S)$ as it accounts for the typical credit portfolio characteristics of the cover pool.[15] The time-T cover pool value is obtained by calculating the risk-neutral expectation of the discounted time-S cash flow,

$$V^{DS}(T) = \mathbb{E}^{\mathbb{Q}}\left[e^{-\int_T^S r(s)ds} \cdot V^{DS}(S) \mid \mathcal{F}_T\right], \quad T \leq S. \qquad (3.7)$$

The determination of the time-T value of the other assets is much simpler,

$$V^{SA}(T) := \tilde{V}^{SA}(T).$$

Here, Sünderhauf [148] argues that a jump-diffusion process is more appropriate since the majority of other assets of a mortgage bank does not have a loan-like payment profile.

Liability modelling. Sünderhauf distinguishes four different liability positions, which in the special case of only one Pfandbrief type simplifies to three positions: a Pfandbrief (PF), other liabilities (SV) and equity (EK). In his model, each liability type $x \in \{PF, SV\}$ is modelled as a risky zero coupon bond with maturity T and nominal N^x. Whether or not the debtors' claims N^x are repaid in full at maturity depends on the then-prevailing value of the bank's assets, with time-T liability cash flows being given by[16]

$$V^{PF}(T) := \min\left(N^{PF}; V^{DS}(T) + V^{SA}(T)\right),$$
$$V^{SV}(T) := \min\left(N^{SV}; \max\left(0; V^{SA}(T) - G^{DS}(T)\right) + E^{DS}(T)\right), \qquad (3.8)$$
$$V^{EK}(T) := \max\left(0; V^{DS}(T) + V^{SA}(T) - N^{PF} - N^{SV}\right),$$

---

[15]By typical credit portfolio characteristics he means a high percentage of loans and bonds with fixed maturity and the maximum payment being capped at the scheduled time-S cash flow, which comprises nominal repayments and interest payments.

[16]For the corresponding formulas in the full model with two Pfandbrief types, see Sünderhauf [148], Chapter 5.1.2.

with

$$E^{DS}(T) := \max(0; V^{DS}(T) - N^{PF}),$$
$$G^{DS}(T) := \max(0; N^{PF} - V^{DS}(T)).$$

The maximum claim of Pfandbrief holders and creditors of other liabilities is given by N^{PF} and N^{SV} respectively. The equations in (3.8) account for the priority claim of Pfandbrief holders on the cover pool and the fact that equity is subordinate, meaning that equity holders only receive payments in cases where both Pfandbrief holders and creditors from other liabilities have been fully repaid. Note that, according to the above equations, Pfandbrief holders also have a priority claim on the other assets, while creditors of other liabilities only receive money once Pfandbrief holders have been fully repaid. This is not in line with practice where remaining claims of Pfandbrief holders, which could not be satisfied by the cover pool, rank pari passu with the claims of other liabilities on the general insolvency estate. In Sünderhauf's one-period model this approximation is necessary because dynamic cover pool management is not possible and the cover requirements can only be reestablished at time T. More specifically, the priority claim of Pfandbrief holders on the other assets approximates the maintenance of cover requirements by the issuer who is required to maintain a certain level of overcollateralization at all times. Under the assumptions made, the time-t value of liability $x \in \{PF, SV, EK\}$ is determined by

$$V^x(t) = \mathbb{E}^{\mathbb{Q}} \left[e^{-\int_t^T r(s)ds} \cdot V^x(T) \,\middle|\, \mathcal{F}_t \right], \quad t \leq T. \tag{3.9}$$

Default specification. In Sünderhauf's model, liabilities can only default at maturity and do so if the time-T value of the bank's assets is not sufficient to repay the outstanding liabilities, $V^{DS}(T) + V^{SA}(T) < N^{PF}(T) + N^{SV}(T)$. In this case, other liabilities experience a loss and there are no payments to equity. Pfandbrief holders, on the contrary, may still be repaid in full. Whether or not there is a loss to Pfandbrief holders depends on the total value of the bank's assets as compared to the nominal of outstanding Pfandbriefe. Pfandbrief holders are repaid in full as long as $V^{DS}(T) + V^{SA}(T) \geq N^{PF}(T)$, otherwise they experience a loss. Figure 3.4 illustrates the time-T liability cash flows arising under the different scenarios.

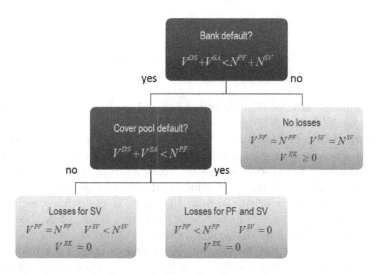

Figure 3.4: Liability cash flows in Sünderhauf's model
$(V^x := V^x(T)$, for $x \in \{DS, SA, PF, SV, EQ\})$.

Implementation.   Due to the complex model setup with two nested structural credit risk models (one for the bank and one for the cover pool), the choice of the stochastic processes for the risk drivers, the integration of term transformation with $S > T$ and the specific liability payment profile there is no analytical solution to Equation (3.9). As pointed out by Sünderhauf [148], pp. 96 and 141, analytical solutions only exist for sub-parts and special cases of his model. This is why he uses Monte Carlo simulation to approximate the present values of liabilities in Equation (3.9), which requires the time-T values of the cover pool and the other assets to be determined in each simulation step, cf. Equation (3.8). For $V^{DS}(T)$ this means that the expectation in Equation (3.7) has to be calculated. Instead of determining this expectation in each simulation path by an additional Monte Carlo simulation, Sünderhauf uses the Least Square Monte Carlo approach by Longstaff and Schwartz (2001) to approximate $V^{DS}(T)$. In his setup, the time-T values of the cover pool's state variable and the short rate are used as explanatory variables and the basis functions are simple polynomials,

$$V^{DS}(T) \approx \beta_0 + \beta_1 \tilde{V}^{DS}(T) + \beta_2 (\tilde{V}^{DS}(T))^2 + \beta_3 r(T) + \beta_4 \left(r(T)\right)^2 . \ (3.10)$$

For a more detailed description of the simulation setup, see Sünderhauf [148], Chapters 4.2 and 5.2. According to Hughes and Werner [90], the performance of the regression equation (3.10) as specified by Sünderhauf is rather weak. The authors show that allowing for slightly more complex basis functions (Laguerre polynomials) and including $\sigma_x(T)$, the time-T value of the stochastic volatility of the interest rate process, as explanatory variable yields considerably better results.

Remark. Sünderhauf calls his model a model for default risk pricing which is misleading to a certain degree. He does not calibrate the model to interest rates and spreads observed in the market and mainly uses it for scenario analyses to determine the impact of certain model parameters, risk drivers and the bank's capital and risk structure on the value of assets and liabilities. The credit spread he calculates for the liabilities is used for the comparison of results, but it is not compared to observed market prices. He even states that, in general, he does not claim to explain the full credit spread observed in the market, but only the default risk part of it, see Sünderhauf [148], p. 77.[17]

3.3.2.2 A Multi-Period Bank Run Model for Liquidity Risk

The dynamic bank run model by Liang et al [114] is a multi-period structural credit risk model for default probability estimation. It explicitly accounts for external funding options and the fact that banks typically finance their assets by a mixture of short-term and long-term debt. The model incorporates multiple rollover dates for short-term debt and distinguishes two different default reasons: overindebtedness and illiquidity. While default due to overindebtedness is caused by a deterioration of asset quality and can occur continuously in time, default due to illiquidity can only occur at a finite number of rollover dates and is triggered by a bank run by short-term creditors in conjunction with not enough funding being available to pay them off.

The bank's balance sheet. Liang et al assume that the asset side of the bank's initial balance sheet consists of risky assets (V) and cash (M). On

---

[17]As discussed in Section 3.2.5, credit spreads are not only driven by default risk but may also be influenced by other factors including liquidity, risk premia and tax effects.

the liability side, there is long-term debt (L) maturing at some time $T > t_0$, and short-term debt (S) which the bank tries to roll over until time T, at discrete rollover dates to be specified in the following. Figure 3.5 shows the bank's initial balance sheet as proposed by Liang et al. Note that the authors do not explicitly mention an equity position on the liability side. For their analyses, this is not necessary as they focus on bank default probability estimation.

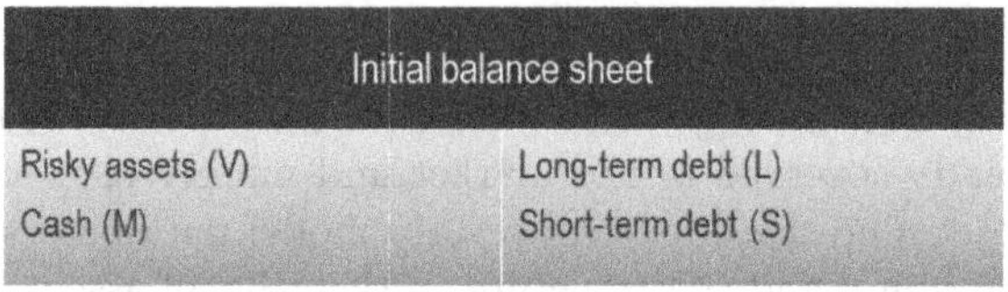

	Initial balance sheet
Risky assets (V)	Long-term debt (L)
Cash (M)	Short-term debt (S)

Figure 3.5: Liang et al's initial balance sheet.

Under the real-world measure $\mathbb{P}$, the bank's risky assets follow a geometric Brownian motion

$$dV(t) = \mu V(t)dt + \sigma V(t)dW(t),$$

with drift $\mu > 0$, volatility $\sigma > 0$ and a standard Brownian motion $(W(t))_{t \geq 0}$.

Cash accrues at the constant risk-free rate r, i.e. $M(t) = M(0) \cdot e^{r \cdot t}$. Furthermore, the bank's long-term debt which matures at time T accrues at a constant annual rate r_L. Unless there is a default, the amount to be repaid at maturity is therefore given by $L(T) = L(0) \cdot e^{r_L \cdot T}$. Liang et al's assumptions with respect to the bank's short-term debt are as follows. There are $n - 1$ rollover dates $t_1, ..., t_{n-1}$ at which short-term creditors decide whether to renew their funding or not, with $t_0 := 0$ and $t_n := T$. For simplicity, it is assumed that the rollover dates are equidistant with $\Delta t := \frac{T}{n}$. Under the assumption that short-term debt has been rolled over at all previous roll-over dates and that it accrues at a constant annual rate r_S, the face value of outstanding short-term debt in t_{i+1} is given by

$$S(t_{i+1}) = S(t_i) \cdot e^{r_S \cdot \Delta t} = S(0) \cdot e^{r_S \cdot (i+1) \cdot \Delta t}. \qquad (3.11)$$

Typically, one would expect $r < r_S < r_L$.

The rollover decision. Regarding the rollover of short-term debt, Liang et al make the following assumptions:

1. Rollover decisions are solely made at rollover dates, and the rollover decision made in t_i only concerns the time period until t_{i+1}.

2. To make the rollover decision, short-term creditors rely on the so-called bank run survival probability λ_i, which represents their subjective beliefs with respect to the behaviour of other short-term creditors,

$$\lambda_i := \min \left(1; \frac{\Psi_i \cdot V(t_i) + M(t_i)}{S(t_i)} \right),$$

   where $\Psi_i \in [0,1]$ is the time-t_i percentage at which cash can be raised by selling risky assets or by posting these assets as collateral.[18] The total amount of cash available to the bank at that time is given by

$$F(t_i) := \Psi_i \cdot V(t_i) + M(t_i).$$

3. The rate r_S at which short-term debt accrues is higher than the corresponding rate r^* in the outside market. In their simulations, the authors set r^* equal to the risk-free rate, i.e. $r_S > r^* := r$. This prevents short-term creditors from investing in the outside market due to better conditions.

4. Once a short-term creditor has decided to withdraw his money at some time t_i where no bank run occurs, he cannot reverse his withdrawal decision and return as a potential lender to the bank at a later time. This is to prevent short-term investors to switch to the outside market and back as they like, which would result in decisions always being made for the next time period only.

5. As long as not too many short-term creditors withdraw their funding, the bank is always able to replace short-term creditors who decide not to roll over by new short-term creditors that can be found in the market. This assumption is needed to ensure that $S(t_i)$ can be calculated as in Equation (3.11).

---

[18]Liang et al motivate the definition of λ_i using standard arguments from game theory. For more details, see their paper.

Collateralized funding. In case a bank run occurs at time $t_R \in \{t_1, ..., t_{n-1}\}$, the bank has to pay off its short-term creditors. Liang et al assume that this can be done by using cash or by pledging risky assets to obtain collateralized funding until time T, at a constant rate $r_C > r_L$.[19] The maximum amount of cash that can be raised at time t_R is given by $F(t_R)$ as defined above. Under the assumption that $M(t_R) \leq S(t_R)$, two different cases can occur:

- $F(t_R) \geq S(t_R)$: The bank uses all its cash $M(t_R)$ to repay short-term creditors and pledges a fraction $\theta_R := \frac{S(t_R)-M(t_R)}{\Psi_R \cdot V(t_R)} \leq 1$ of its risky assets at the rate Ψ_R. The total amount of cash available is then

$$\theta_R \cdot \Psi_R \cdot V(t_R) + M(t_R) = S(t_R),$$

  which means that short-term debt can be fully repaid. After a bank run has occurred, there is no more cash left on the bank's balance sheet and the amount of risky assets is reduced to $\tilde{V}(t_R) := (1 - \theta_R) \cdot V(t_R)$. Long-term debt remains unchanged and a new position appears on the balance sheet: collateralized debt with maturity T and corresponding time-T payment given by

$$W(T) := \theta_R \cdot \Psi_R \cdot V(t_R) \cdot e^{r_C \cdot (T-t_R)} = (S(t_R) - M(t_R)) \cdot e^{r_C \cdot (T-t_R)}.$$

  If the bank is not able to make this payment at time T, the lender keeps the collateral which then has a value of $\theta_R \cdot V(T)$. The counterpart of the collateralized funding transaction is therefore exposed to market risk. He has provided a cash amount of $\theta_R \cdot V(t_R) \cdot \Psi_R$ to the bank and expects a repayment of $W(T)$ at time T. Should the bank default at that time and the asset value has deteriorated such that $\theta_R \cdot V(T) < W(T)$, he does not get the full expected payment.[20]

- $F(t_R) < S(t_R)$: The bank is not able to raise sufficient cash to repay $S(t_R)$ and defaults. In this case Liang et al assume that neither short-term nor long-term creditors receive any payments (zero recovery).[21]

---

[19]The rate at which the collateralized funding can be obtained reflects the return required by market participants to lend money to a bank which is already in troubles. As both types of debt mature at time T, it can be reasonably assumed that $r_C > r_L$.

[20]This is why in practice secured funding transactions often require the borrower to post additional collateral once the value of the collateral deteriorates.

[21]As the authors' main interest is estimating default probabilities, recovery does not play a role for them.

Depending on the bank's ability to raise collateralized funding, a bank run does not necessarily result in a default. Figure 3.6 shows the bank's balance sheet after a bank run in case it survives.

In the following, we denote by $t_R \in \{t_1, ..., t_{n-1}, \infty\}$ the first time of a bank run. As Liang et al assume that a bank run can only occur once during the bank's lifetime, it is also the only time of a bank run, and $t_R = \infty$ is to be interpreted as no bank run occurring at the specified rollover dates.

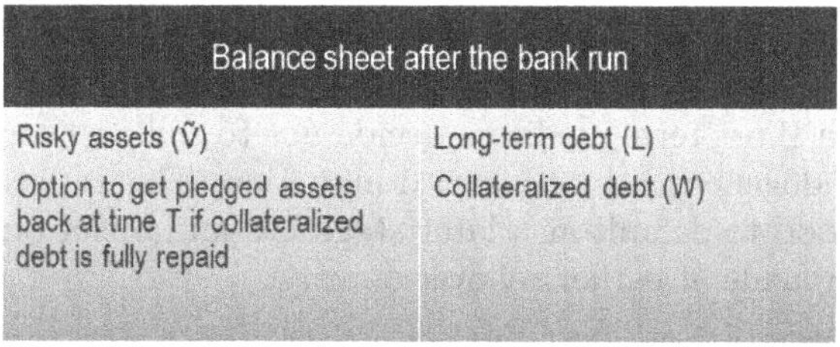

Balance sheet after the bank run	
Risky assets ($\tilde{V}$)	Long-term debt (L)
Option to get pledged assets back at time T if collateralized debt is fully repaid	Collateralized debt (W)

Figure 3.6: The bank's balance after a bank run in case it survives.

Specification of the insolvency barrier.   Liang et al use an exogenous time-dependent insolvency barrier similar to the one in Black and Cox [19], see Equation (3.2). As such a barrier depends on the structure of the bank's balance sheet, two cases are distinguished:

- Until a bank run has occurred, $t_0 \le t \le t_R$, the bank has short-term and long-term debt outstanding, and the bank's assets consist of cash and risky assets. The corresponding insolvency barrier is defined by

$$V_B^S(t) := \alpha \cdot P(t,T) \cdot \left[S(0) \cdot e^{r_S \cdot T} + L(0) \cdot e^{r_L \cdot T} - M(0) \cdot e^{r \cdot T} \right],$$

  with $\alpha \in [0,1]$ being a safety covenant that specifies how much asset value must at least be available for the compensation of creditors upon default and $P(t,T) := e^{-r \cdot (T-t)}$.

- Once a bank run has occurred, $t_R < t \le T$, the situation is different as there is no more short-term debt outstanding. Instead, there is now collateralized debt. As the bank's risky assets have been reduced, the resulting insolvency barrier is given by

$$\tilde{V}_B^S(t) := \alpha \cdot P(t, T)$$
$$\cdot \left[\left(S(t_R) - M(0) \cdot e^{r \cdot t_R} \right) \cdot e^{r_C \cdot (T - t_R)} + L(0) \cdot e^{r_L \cdot T} \right].$$

Specification of the bank run barrier. At each rollover date, short-term creditors decide whether to roll over their debt or to run on the bank. They make their decision based on the expected return resulting from these two options. The bank run barrier $V_B^R(t_i)$ is derived as the critical asset value at which short-term creditors decide not to roll over any more. It is determined endogenously, by backward induction. This is necessary as the bank's likelihood to default at a later stage has an impact on the rollover decision being made at earlier rollover dates.

At time t_{n-1}, the short-term creditors' considerations are as follows:

- The expected return from running the bank in t_{n-1} and investing the money in the outside market until time t_n is obtained from

$$e^{R_{n-1}^* \cdot \Delta t} = e^{r^* \cdot \Delta t} \cdot \lambda_{n-1}, \qquad (3.12)$$

  with r^* being the rate at which the investment accrues in the outside market. Equation (3.12) reflects the fact that short-term investors only get their money back if the bank survives the run, which is expected to occur with probability λ_{n-1}. Otherwise they get nothing (zero recovery). A short-term creditor who runs at time t_{n-1} is not affected by a potential future bank insolvency at times $s > t_{n-1}$.

- The expected return from rolling over from time t_{n-1} to t_n is derived from

$$e^{R_{n-1}^S \cdot \Delta t} = e^{r_S \cdot \Delta t} \cdot q_{n-1,n} \cdot \lambda_{n-1}, \qquad (3.13)$$

  with $q_{n-1,n}$ being the conditional survival probability from insolvency within the time period $(t_{n-1}, t_n]$ given $V(t_{n-1})$,

$$q_{n-1,n} := \mathbb{P}\left[\inf_{t_{n-1} < s \leq t_n} \left(V(s) - V_B^S(s) \right) \geq 0 \Big| V(t_{n-1}) \right].$$

Equation (3.13) reflects the fact that short-term creditors only get their money back in t_n if the bank does not default due to insolvency in $(t_{i-1}, t_i]$ or due to a bank run in t_{n-1}. If the bank defaults for one of these two reasons, short-term creditors get nothing (zero recovery).

- If the bank defaults due to insolvency at time t_{n-1}, there is no rollover decision to be made, which is why Equations (3.12) and (3.13) are not multiplied by the corresponding survival probability from insolvency at time t_{n-1}.

- R_{n-1}^S and R_{n-1}^* can be backed out from Equations (3.12) and (3.13). As it is assumed that short-term creditors maximize their expected return over $[0, T]$, they run on the bank if R_{n-1}^S is below R_{n-1}^*. The bank run barrier $V_B^R(t_{n-1})$ is therefore determined by setting the right hand side of Equation (3.12) equal to the right hand side of Equation (3.13) and solving for the critical asset value $V(t_{n-1})$, which appears in the definition of $q_{n-1,n}$,

$$e^{r^* \cdot \Delta t} \cdot \lambda_{n-1} \stackrel{!}{=} e^{r_S \cdot \Delta t} \cdot q_{n-1,n} \cdot \lambda_{n-1}.$$

By doing so, λ_{n-1} cancels out, i.e. the bank run barrier at time t_{n-1} is not impacted by the survival probability λ_{n-1}.

- The maximum rate short-term creditors can earn from time t_{n-1} to time t_n is given by $\max\left(R_{n-1}^*; R_{n-1}^S\right)$.

Proceeding backward in time, short-term creditors compare their return over the period from time t_i to t_n for making their decision in t_i, for $i = n-2, ..., 1$.

- The expected return from running the bank in t_i and investing the money in the outside market until time t_n is given by

$$e^{R_i^* \cdot (n-i) \cdot \Delta t} = e^{r^* \cdot (n-i) \cdot \Delta t} \cdot \lambda_i. \tag{3.14}$$

Again, short-term debt is only paid back if the bank survives the bank run, which is expected to occur with probability λ_i.

- The expected return from rolling over at time t_i is given by

$$e^{R_i^S \cdot (n-i) \cdot \Delta t} = k_i \cdot e^{r_S \cdot \Delta t} \cdot q_{i,i+1} \cdot \lambda_i, \tag{3.15}$$

with $q_{i,i+1}$ being the conditional survival probability from insolvency within the time period $(t_i, t_{i+1}]$ given $V(t_i)$,

$$q_{i,i+1} := \mathbb{P}\left[\inf_{t_i < s \leq t_{i+1}} \left(V(s) - V_B^S(s) \right) \geq 0 \,\middle|\, V(t_i) \right],$$

and k_i being the expected return from future periods conditional on $V(t_i)$,

$$k_i := \mathbb{E}\left[e^{\max(R_{i+1}^*; R_{i+1}^S) \cdot (n-i-1) \cdot \Delta t} \,\middle|\, V(t_i) \right].$$

Note that k_i depends on the survival probabilities $\lambda_{i+1}, ..., \lambda_{n-1}$ from potential future bank runs. This implies that short-term creditors which roll over their debt do not only have to worry about bank runs at time t_i but also about bank runs at future rollover dates.

- R_i^S and R_i^* can now be backed out from Equations (3.14) and (3.15). Again, the bank run barrier $V_B^R(t_i)$ is determined by setting the right hand side of (3.14) equal to the right hand side of Equation (3.15) and solving for the critical asset value $V(t_i)$.

- The maximum rate the short-term creditor can earn from time t_i to time t_n is now given by $\max\left(R_i^*; R_i^S \right)$.

Liang et al prove that, given their assumptions, the bank run barrier can be determined recursively and that it corresponds to the unique solution β_i of

$$e^{r^* \cdot (n-i) \cdot \Delta t} \overset{!}{=} \mathbb{E}\left[e^{\max(R_{i+1}^*; R_{i+1}^S) \cdot (n-i-1) \cdot \Delta t} \,\middle|\, V(t_i) = \beta_i \right]$$
$$\cdot e^{r_S \cdot \Delta t} \cdot \mathbb{P}\left[\inf_{t_i < s \leq t_{i+1}} \left(V(s) - V_B^S(s) \right) \geq 0 \,\middle|\, V(t_i) = \beta_i \right]$$

for each rollover date $t_i \in \{t_1, ..., t_{n-1}\}$. They also show that $V_B^R(t_i) \geq V_B^S(t_i)$ for fixed t_i.

Specification of the illiquidity barrier. In Liang et al's model, the bank defaults due to illiquidity if there is a bank run and the bank is not able to raise enough funds to pay off the short-term creditors. As $F(t_i) < S(t_i)$ is equivalent to $V(t_i) < \frac{S(t_i) - M(t_i)}{\Psi_i}$, the illiquidity barrier $V_B^L(t_i)$ is given by

$$V_B^L(t_i) := \min\left(V_B^R(t_i); \frac{S(t_i) - M(t_i)}{\Psi_i} \right) \leq V_B^R(t_i).$$

This threshold is only specified for the given rollover dates, meaning that default due to illiquidity can only occur at a finite number of dates. Once a bank run has occurred, the illiquidity barrier becomes irrelevant. Figure 2 in Liang et al [114] illustrates the event of default due to illiquidity in Liang et al's model.

Potential scenarios at time t. For a specific rollover date t_i and under the assumption that no bank run has occurred prior to that date, Liang et al distinguish the following four scenarios:

- **Case i:** If $V(t_i) \leq V_B^S(t_i)$, the bank defaults due to insolvency.

- **Case ii:** If $V_B^S(t_i) < V(t_i) < V_B^L(t_i) \leq V_B^R(t_i)$, the bank is solvent, but there is a bank run and the bank is not able to raise sufficient cash to pay off short-term creditors. This results in default due to illiquidity.[22]

- **Case iii:** If $V_B^L(t_i) \leq V(t_i) \leq V_B^R(t_i)$, there is a bank run but the bank is able to pay off short-term creditors. Here, two sub cases are distinguished:

  - **Case iiia:** In case $V(t) \leq \tilde{V}_B^S(t)$ for some $t \in (t_i, T]$, the bank defaults. Liang et all refer to this event as default due to illiquidity and motivate it by arguing that it was actually caused by the bank run at time t_i.[23]

  - **Case iiib:** In case $V(t) > \tilde{V}_B^S(t)$ for all $t \in (t_i, T]$, the bank remains solvent until time T.

In the following, we provide a more general description of the potential scenarios for an arbitrary time t, under the assumption that no default has occurred prior to that time. We do so by slightly adjusting the definition of illiquidity, which is due to the aforementioned peculiarity. This generalized scenario specification will be used as a reference later on. All in all, we distinguish eight scenarios. If no bank run has occurred prior to time $t < t_R$

---

[22] Note that this is a slightly adjusted version of the specification given by Liang et al. The authors assume that default due to illiquidity occurs once $V_B^S(t_i) < V(t_i) \leq V_B^L(t_i)$, which includes the case were $V_B^S(t_i) < V(t_i) = V_B^L(t_i)$, implying $F(t_i) = S(t_i)$, i.e. short-term creditors could actually still be repaid. Our modification also concerns Case iii below which now includes $V_B^L(t_i) = V(t_i)$.

[23] An alternative point of view could be to interpret this as a default due to overindebtedness, triggered by the bank's asset value falling below the solvency boundary $\tilde{V}_B^S(t)$.

and t is not a rollover date, a bank run is not possible per definition and two different situations need to be distinguished:

- **Case 1:** If $V(t) \leq V_B^S(t)$, the bank defaults due to insolvency.

- **Case 2:** If $V(t) > V_B^S(t)$, the bank survives.

If there was no bank run prior to time t and t is a rollover date, what happens depends on whether or not a bank run occurs. In case there is no bank run, there is only one possible scenario:

- **Case 3:** As $V(t) > V_B^R(t)$ implies $V(t) > V_B^S(t)$ and $V(t) > V_B^L(t)$, the bank survives.

In case of a bank run, $V(t) \leq V_B^R(t)$, three sub-scenarios need to be distinguished:

- **Case 4:** If the bank is still solvent, $V(t) > V_B^S(t)$, and able to raise enough cash to pay off short-term creditors, $V(t) \geq V_B^L(t)$, it survives.

- **Case 5:** If the bank is solvent but illiquid, implying that $V(t) > V_B^S(t)$ and $V(t) < V_B^L(t) \leq V_B^R(t)$, the bank does not survive the bank run as it is not able to raise enough cash to pay off short-term creditors.

- **Case 6:** Once the insolvency barrier is hit, $V(t) \leq V_B^S(t) \leq V_B^R(t)$, the bank defaults due to insolvency.

Finally, if a bank run has occurred already prior to time t, bank default solely depends on whether the solvency barrier $\tilde{V}_B^S(t)$ – which is now different from the one in the previous cases – is hit or not and there are two potential scenarios to be distinguished:

- **Case 7** If $V(t) \leq \tilde{V}_B^S(t)$, the bank defaults.

- **Case 8** If $V(t) > \tilde{V}_B^S(t)$, the bank survives.

Figure 3.7 summarizes the discussed scenarios. It can be easily seen that, apart from the above discussed adjustments, the distinction of these eight scenarios is a generalization of the distinction made by Liang et al. As the authors only consider the case of a bank run at a rollover date t_i, Cases 1 to 3 do not occur in their distinction. Furthermore, their Case i corresponds to our Case 6, and their Case ii corresponds to our Case 5. At the exact time of the bank run, their Cases iiia and iiib both result in our case 4. After the bank run, our Cases 7 and 8 describe the fate of the bank, which corresponds to their Cases iiia and iiib, respectively.

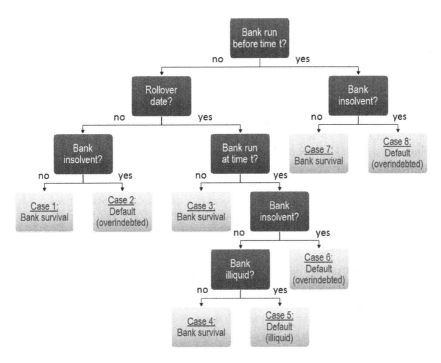

Figure 3.7: Generalized version of potential scenarios in Liang et al's model.

Implementation. Liang et al implement their model in a binomial tree setting and solve it numerically. Taking into account additional stochastic risk drivers such as several processes for the creditworthiness of risky assets would therefore result in a multi-dimensional binomial tree setup and complicate the modelling considerably.

3.4 Summary and Conclusion

In this chapter we gave an overview of the existing credit risk literature with particular focus on bank defaults and Pfandbrief modelling requirements. Out of the two main credit risk approaches, the structural approach turned out to be the natural choice as it allows for a link between the market environment and potential default events. The Merton model which forms the basis for structural credit risk models is based on quite restrictive assumptions and therefore not suitable for our purpose. This is why several features and model extensions as found in the literature were discussed. With respect to the specification of the default barrier, we found exogenous barriers to be preferable as they are more intuitive and easier to derive. In this setup, default can be interpreted as some kind of overindebtedness. As banks typically rely on debt-rollover and illiquidity may be a potential default reason, we reviewed different funding assumptions and their practical relevance as well as related model implications. We also discussed the application of structural credit risk models to bank default modelling in general and found that there are only few models which are particularly designed for banks. Even though we could not find any structural credit risk model which entirely fits our purpose, there were two models which caught our attention as they exhibit interesting features in the context of Pfandbrief modelling.

Sünderhauf's mortgage bank model takes into account the specific structure of a Pfandbrief bank's balance sheet and distinguishes between cover pool assets and other assets as well as between Pfandbriefe and other liabilities. Due to his one-period setup, cash flows are also implicitly modelled. He further distinguishes between issuer and cover pool default and establishes a link between the market environment and the corresponding default events. Default is specified with the help of exogenous default barriers while recovery rates are determined endogenously. In his model, Sünderhauf allows for stochastic interest rates, incorporates jumps in the asset value process and considers the impact of these risk drivers on the bank's and the cover pool's creditworthiness. The Pfandbrief's dual protection mechanism, the legal cover requirements and the Pfandbrief-specific priority of payments in case of default are also accounted for. As the model is a one-period model only, dynamic cover pool management and funding and reinvestment activities cannot be modelled. This essentially ignores all aspects related to bank liquidity, debt rollover, access to external financing and default due to illiquidity. In addition, the balance sheet structure in his model is too simplistic to allow for realistic Pfandbrief modelling. Bankruptcy and liquidation costs are also not considered.

The multi-period bank run model by Liang et al incorporates both default due to overindebtedness and default to illiquidity. To specify the default events, the model relies on three different kinds of thresholds: an insolvency barrier, a bank run barrier and an illiquidity barrier. Default due to illiquidity is triggered by a bank run in conjunction with not enough funding being available to pay off short-term creditors. The model explicitly accounts for debt rollover, external funding options and the inability to obtain funding under certain conditions. It also establishes a link between the market environment and bank performance. The model does, however, not account for any Pfandbrief-specific features which include the Pfandbrief mechanics (i.e. the dual protection mechanism, the legal cover requirements, the ring-fencing upon issuer default and the priority of payments in case of default), dynamic cover pool management and a distinction of the events of issuer and cover pool default. Moreover, it relies on a very simplistic debt structure with only one type of long-term debt and one type of short-term debt and makes further simplifying assumptions such as constant interest rates and risky assets following a lognormal diffusion process. The recovery rate is exogenously given and set to 0, which implies bankruptcy costs of 100%.

Table 3.2 shows our required modelling features as derived in Section 2.3 and relates them to the two above discussed models. All in all, we conclude that we need to develop a new Pfandbrief model to incorporate all these features. As we will see in the next chapter, our model is strongly inspired by Sünderhauf and Liang et al.

Table 3.2: Required features for Pfandbrief modelling in the models by Sünderhauf (SH) and Liang et al (LI).

Model component	Feature	SH	LI
General	Multi-period time setting	-	✓
Default modelling	Distinction of issuer and cover pool default	✓	-
	Distinction of overindebtedness and illiquidity	-	✓
	Link between market environment and default	✓	✓
Balance sheet	Adequate representation of risk profile	-	-
	Distinction of Pfandbrief business	✓	-
	Modelling of asset present values *and* cash flows	✓	✓
Market environment	Stochastic risk drivers	✓	✓
	Impact on asset present values and cash flows	✓	✓
ALM	Active cover pool management	-	-
	Funding and reinvestment activities	-	✓
Pfandbrief mechanics	Dual protection mechanism	✓	-
	Maintenance of legal cover requirements	✓	-
	Ring-fencing upon issuer default	✓	-
	Priority of payments in case of default	✓	-

4 The Pfandbrief Model

In this chapter we develop a multi-period Pfandbrief model which incorporates the required modelling features as derived in Section 2.3. Due to the complexity of the modelling, a stochastic simulation-based framework is our method of choice. The simulation starts in $t_0 := 0$ and the time setting is discrete and finite,

$$\mathcal{T}_d := \{t_0, ..., t_S\}, \quad S \in \mathbb{N},$$

where $t_{i+1} := t_i + \Delta$ for $i = 0, ..., S - 1$, with $\Delta := \frac{0.5}{k}$ and $k \in \mathbb{N}$.[1]

Throughout this work we assume that we are dealing with a frictionless financial market. The uncertainty in this market is modelled by a filtered probability space $(\Omega, \mathcal{F}, \mathbb{P})$ with Ω being the set of possible states of nature, $\mathcal{F} := (\mathcal{F}_t)_{t \in \mathcal{T}_d}$ the natural filtration and $\mathbb{P}$ the real-world probability measure. We further assume that the underlying financial market is arbitrage-free and $\mathbb{Q}$ denotes some equivalent martingale measure. All relevant events happen at the discrete time steps $t \in \mathcal{T}_d$ which are further divided into three sub-time steps, see Table 4.1. At the *beginning of the period* (bop), the impact of market movements on the bank's balance sheet is determined. This is done by simulating the time-t values of the stochastic risk drivers and repricing all relevant asset positions. If necessary, liability positions are also adjusted. *Mid of period* (mop), asset cash inflows are collected and the current funding need is determined, which involves the identification of due liability payments and the performance of matching cover calculations. Then, it is checked whether an event of bank or cover pool default occurs. At the *end of the period* (eop), asset liability management activities take place. This includes funding and reinvestment decisions, the maintenance of overcollateralization, potential asset liquidations and liability payments. Finally, the balance sheet is updated accordingly.

Our simulation model distinguishes nine Pfandbrief scenarios which trigger different liability cash flows and asset liability management related activities.

---

[1] A simulation time step size $\Delta \leq 0.5$ is necessary for a meaningful modelling of the 180-day liquidity buffer.

© Springer Fachmedien Wiesbaden GmbH, part of Springer Nature 2018
M. Spangler, *Modelling German Covered Bonds*, Mathematische Optimierung
und Wirtschaftsmathematik | Mathematical Optimization and Economathematics,
https://doi.org/10.1007/978-3-658-23915-2_4

Table 4.1: The three simulation sub-time steps.

Sub-time step	Description
bop	Impact of market movements on the balance sheet
	- Simulation of stochastic risk drivers
	- Update of asset present values
	- Adjustment of liability positions
mop	Liquidity situation and potential default events
	- Collection of asset cash inflows
	- Determination of funding need
	- Check for default events
eop	Asset liability management
	- Funding and reinvestment activities
	- Maintenance of overcollateralization
	- Potential asset liquidations
	- Liability payments
	- Balance sheet update

As this distinction forms the basis of our modelling approach, Section 4.1 starts with a description of these scenarios. In Section 4.2 we define the bank's balance sheet, which can be considered to be a modified extension of the balance sheet used by Sünderhauf [148]. It consists of several classes of assets and liabilities with different nominals, maturities and riskiness and therefore allows for an explicit distinction of the balance sheet positions which are related to the bank's Pfandbrief business, i.e. the cover pool and the outstanding Pfandbriefe. The asset performance depends on stochastic risk drivers: the risk-free interest rate and state variables for the assets' creditworthiness. Section 4.3 is dedicated to the market environment. The dynamics of the stochastic risk drivers are specified and their impact on asset present values and cash flows is determined. Section 4.4 addresses the liquidity situation of the bank and the cover pool and specifies the conditions under which funding is available in our model. In Section 4.5 we then define the events of bank and cover pool default and the priority of payments in case of a liquidation event. Motivated by Liang et al [114] we distinguish two different default triggering events: overindebtedness and illiquidity. Section 4.6 deals with the asset liability management strategies of the bank and the cover pool administrator, which include funding and reinvestment strategies, the maintenance of overcollateralization and liability payments. Having specified all model components, we derive the liabilities' default parameters in Section 4.7.

4.1 Distinction of Nine Pfandbrief Scenarios

Depending on the occurrence of the events of bank and cover pool default and the simulation time t, we distinguish nine Pfandbrief scenarios as shown in Figure 4.1. To illustrate the broad idea of our modelling approach, we shortly describe these scenarios and their implications for liability cash flows and asset liability management in the following. All related definitions and assumptions are specified in detail in the subsequent sections. With $T_{\max} \in \mathcal{T}_d$ denoting the maturity of the longest Pfandbrief outstanding at time t_0, our distinction of the nine scenarios is as follows.

Scenario S1: "Business as usual". As long as the bank has not defaulted and $t < T_{\max}$, the Pfandbrief holders and the creditors of other liabilities receive scheduled payments.[2] The bank also performs asset liability management activities, which include the maintenance of overcollateralization. In this scenario, there is no distinction between cash inflows from the cover pool and the bank's other assets. All cash inflows are equally used to fulfil payment obligations and to reestablish cover requirements, i.e. there is no dedicated one-to-one relationship between the cover pool and outstanding Pfandbriefe.

Scenario S2: "Creation of a standalone cover pool". If the bank defaults at time $t < T_{\max}$ but the cover pool does not, one needs to distinguish between the fate of Pfandbriefe and the fate of other liabilities. The bank's non-cover pool assets, which belong to the bank's general insolvency estate, are liquidated and the proceeds are used to repay the creditors of other liabilities. If these proceeds are sufficient to satisfy the liquidation claims of all outstanding other liabilities, shareholders receive the remainder. The cover pool is segregated and survives the event of bank default. The cover pool administrator, who is appointed to take care of the cover pool, uses the cover pool's cash inflows to continue scheduled payments to Pfandbrief holders. He still performs certain asset liability management actions, but the cover requirements are not maintained any more.

Scenario S3: "Cover pool liquidation at bank default". In case of a simultaneous bank and cover pool default at time $t < T_{\max}$, all assets are liquidated. Pfandbrief holders have a priority claim on the proceeds from cover pool liquidation and rank pari passu with the creditors of other liabilities with respect to the bank's general insolvency estate. If the liquidation value of the cover pool exceeds the value of the liquidation claims of Pfandbrief

---

[2]By assumption, there are no dividend payments to shareholders in our setup.

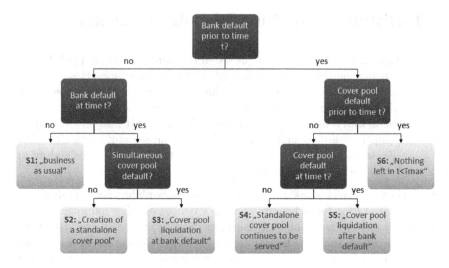

(a) Pfandbrief scenarios in $t < T_{\max}$.

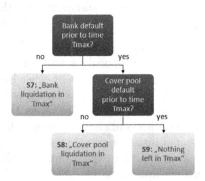

(b) Pfandbrief scenarios in $t = T_{\max}$.

Figure 4.1: Definition of the nine Pfandbrief scenarios.

holders, the remainder can be used to repay the creditors of other liabilities. Shareholders only receive payments after all liabilities have been fully repaid.

Scenario S4: "Standalone cover pool continues to be served". If the standalone cover pool does not default at time $t < T_{max}$, the cover pool administrator continues managing the cover pool and Pfandbrief holders receive scheduled payments. Cover requirements are not maintained any more. As the bank has already defaulted at some previous time, there are no more payments to the creditors of other liabilities or to shareholders.

Scenario S5: "Cover pool liquidation after bank default". In case the standalone cover pool defaults at time $t < T_{max}$, the cover pool is liquidated and the liquidation proceeds are used to repay Pfandbrief holders. If the liquidation proceeds exceed the liquidation claims of Pfandbrief holders, they are used to reduce the losses incurred by the creditors of other liabilities at the (previous) time of bank default. Again, shareholders only receive payments after all liabilities have been fully repaid. In our model, Pfandbrief holders do not have a claim on the general insolvency estate in this scenario.

Scenario S6: "Nothing left in $t < T_{max}$". If both the bank and the cover pool have already defaulted prior to time $t < T_{max}$, no more business activities take place.

Scenario S7: "Bank liquidation in T_{max}". In case the bank has not defaulted prior to time T_{max}, all assets on the bank's balance sheet are liquidated in T_{max} and the proceeds are used to repay the Pfandbrief holders and the creditors of other liabilities. If the liquidation proceeds are sufficient to satisfy all these claims, shareholders receive the remainder. The priority of payments is the same as in scenario S3, but in our model liquidation in T_{max} occurs regardless of a bank default event.

Scenario S8: "Cover pool liquidation in T_{max}". If the bank has already defaulted prior to time T_{max} but the cover pool has not, the cover pool is liquidated and liquidation proceeds are used to repay Pfandbrief holders. If the liquidation proceeds exceed these liquidation claims, they are used to reduce losses incurred by the creditors of other liabilities at the (previous) time of bank default. Shareholders only receive payments after all Pfandbriefe and other liabilities have been fully repaid. The priority of payments is the same as in scenario S5, but liquidation in T_{max} occurs regardless of a cover pool default event.

Scenario S9: "Nothing left in T_{max}". In case both the bank and the cover pool have already defaulted prior to time T_{max}, no more business activities take place. Scenario S9 is basically the same as scenario S6. The distinction between $t = T_{\max}$ and $t < T_{\max}$ is only needed for technical purposes.

The scenarios S1 to S9 approximate the Pfandbrief mechanics as described in Chapter 2 with a reasonable degree of accuracy. As long as the issuer has not defaulted, cover requirements are maintained and scheduled payments to liabilities are made. Upon bank default, all outstanding other liabilities become due and payable immediately and their claims are satisfied as far as possible with the liquidation proceeds of the bank's non-cover pool assets, which form the general insolvency estate. The cover pool, on the contrary, is segregated and Pfandbrief holders continue to receive scheduled payments as long as the cover pool remains solvent. Pfandbriefe become due and payable only upon cover pool default and are repaid out of the liquidation proceeds of the cover pool. If these are not sufficient, Pfandbrief holders also have a claim on the bank's general insolvency estate. In our model, this claim is only enforceable in case of a simultaneous default of the bank and the cover pool (S3), but not if the cover pool defaults at some later time (S5). Cover pool liquidation proceeds which are not needed to repay Pfandbrief holders are released to the bank's general insolvency estate. Table 4.2 summarizes the main characteristics of the nine scenarios. By defining these scenarios, we set the foundation for our model to be able to incorporate the required features for Pfandbrief modelling as found in Section 2.3.

Table 4.2: Summary of the nine Pfandbrief scenarios.

	S1	S2	S3	S4	S5	S6	S7	S8	S9
Asset liability management									
- Cover requirements	✓	-	-	-	-	-	-	-	-
- Other ALM activities	✓	✓	-	✓	-	-	-	-	-
Pfandbrief payments									
- Scheduled payments	✓	✓	-	✓	-	-	-	-	-
- Liquidation payments	-	-	✓	-	✓	-	✓	✓	-
Payments to other liabilities									
- Scheduled payments	✓	-	-	-	-	-	-	-	-
- Liquidation payments	-	✓	✓	-	(✓)	-	✓	(✓)	-
Balance sheet liquidation									
- Non-cover pool assets	-	✓	✓	-	-	-	✓	-	-
- Cover pool	-	-	✓	-	✓	-	✓	✓	-

4.2 The Bank's Balance Sheet

In the following, we specify the Pfandbrief bank's balance sheet. We do so by distinguishing three kinds of assets, strategic cover pool assets (CPS), liquid cover pool assets (CPL) and other assets (OA), and two kinds of liabilities, Pfandbriefe (PB) and other liabilities (OL). Strategic cover pool assets represent the bank's Pfandbrief business and are the basis for matching cover calculations while liquid cover pool assets, modelled as a risk-free cash account, are solely needed for liquidity steering in the context of cover requirements.[3] Except for liquid cover pool assets, all above positions are split into several sub-positions, with different nominals, maturities and riskiness. Figure 4.2 illustrates the basic structure of our balance sheet, excluding the above mentioned sub-positions but including the residual equity position (EQ).

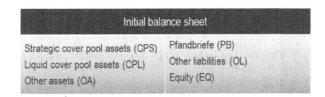

Figure 4.2: The bank's initial balance sheet.

Before we define the bank's initial balance sheet in full detail, we introduce some notation.

Risk-free zero coupon bond. A risk-free zero coupon bond with maturity $T \in \mathcal{T}_d$ is a contract which guarantees a payoff of one unit of currency at time T with no intermediate payments. The present value of this contract at time $t \leq T$, $t \in \mathcal{T}_d$, is denoted by $P(t,T)$, with $P(t,T) > 0$ and $P(T,T) = 1$.

Risk-free spot rate. The risk-free spot rate prevailing at time $t \in \mathcal{T}_d$ for the maturity $T \in \mathcal{T}_d$ is defined by

$$r(t,T) := -\frac{1}{T-t} \ln P(t,T), \quad t < T. \tag{4.1}$$

---

[3]For more details, see Sections 4.4.1 and 4.6.3 below.

Risky zero coupon bond. A risky zero coupon bond with maturity $T \in \mathcal{T}_d$ is a contract which pays one unit of currency at time T if the reference name has not defaulted until then and a random recovery value $\delta \in [0,1]$ otherwise. The present value of this contract at time $t \leq T$, $t \in \mathcal{T}_d$, is denoted by $\tilde{P}(t,T)$, with $\tilde{P}(t,T) \geq 0$ and $\tilde{P}(T,T) = \mathbb{I}_{\{\tau_R > T\}} \cdot 1 + \mathbb{I}_{\{\tau_R \leq T\}} \cdot \delta$, where τ_R is the default time of the reference name. The bond's lifetime default probability is

$$\pi(T) := \mathbb{P}\left[\tau_R \leq T\right],$$

and the associated expected lifetime loss given default is

$$L(T) := \begin{cases} 1 - \mathbb{E}^{\mathbb{P}}[\tilde{P}(T,T)|\tau_R \leq T] = 1 - \mathbb{E}^{\mathbb{P}}[\delta|\tau_R \leq T], & \text{if } \pi(T) > 0, \\ 0, & \text{otherwise.} \end{cases}$$

Spread. At time $t \in \mathcal{T}_d$, the spread of a risky zero coupon bond with maturity $T \in \mathcal{T}_d$ and present value $\tilde{P}(t,T) > 0$ is defined by

$$s(t,T) := -\frac{1}{T-t}(\ln \tilde{P}(t,T) + r(t,T) \cdot (T-t)), \quad t < T.$$

As $e^{r(t,T) \cdot (T-t)} \cdot P(t,T) = 1$, the risk-free spot rate $r(t,T)$ is the constant rate at which an amount $P(t,T)$, invested at time t, accrues until time T to yield a payoff of 1. Similarly, $e^{(r(t,T)+s(t,T)) \cdot (T-t)} \cdot \tilde{P}(t,T) = 1$, i.e. the spread can be interpreted as an additional premium above the risk-free spot rate which is required for a risky investment. The lifetime default probability $\pi(T)$ reflects the likelihood that the risky zero coupon bond is not fully repaid at maturity and the expected lifetime loss given default $L(T)$ quantifies the expected severity of the associated loss.

With respect to the bank's initial balance sheet, we now make the following assumption:

Assumption 1 (Initial balance sheet)
At time t_0, the asset side of the bank's balance sheet consists of

- *$n_{CPS} \in \mathbb{N}^+$ strategic cover pool assets with nominals $N_i^{CPS}(t_0) \in \mathbb{R}^+$, $i = 1, ..., n_{CPS}$, and maturities $T_i^{CPS} \in \mathcal{T}_d$, where $t_0 < T_1^{CPS} \leq ... \leq T_{n_{CPS}}^{CPS}$, modelled as positions in risky zero coupon bonds $\tilde{P}_i^{CPS}(t, T_i^{CPS})$ which can only default at maturity and do so* with a probability of $\pi_i^{CPS} \in (0,1)$ and an expected loss given default of $L_i^{CPS} \in (0,1)$,

- $n_{OA} \in \mathbb{N}^+$ other assets with nominals $N_j^{OA}(t_0) \in \mathbb{R}^+$, $j = 1, ..., n_{OA}$, and maturities $T_j^{OA} \in \mathcal{T}_d$, where $t_0 < T_1^{OA} \leq ... \leq T_{n_{OA}}^{OA}$, modelled as positions in risky zero coupon bonds $\tilde{P}_j^{OA}(t, T_j^{OA})$ which can only default at maturity and do so with a probability of $\pi_j^{OA} \in (0,1)$ and an expected loss given default of $L_j^{OA} \in (0,1)$, and

- a risk-free cash account with nominal $N^{CPL}(t_0) \in \mathbb{R}_0^+$, which accrues at the risk-free spot rate until the next time step,

and the bank's liabilities are given by

- $n_{PB} \in \mathbb{N}^+$ Pfandbriefe with nominals $N_k^{PB}(t_0) \in \mathbb{R}^+$, $k = 1, ..., n_{PB}$, and maturities $T_k^{PB} \in \mathcal{T}_d$, where $t_0 < T_1^{PB} \leq ... \leq T_{n_{PB}}^{PB}$, modelled as positions in risky zero coupon bonds $\tilde{P}_k^{PB}(t, T_k^{PB})$,

- $n_{OL} \in \mathbb{N}^+$ other liabilities with nominals $N_l^{OL}(t_0) \in \mathbb{R}^+$, $l = 1, ..., n_{OL}$, and maturities $T_l^{OL} \in \mathcal{T}_d$, where $t_0 < T_1^{OL} \leq ... \leq T_{n_{OL}}^{OL}$, modelled as positions in risky zero coupon bonds $\tilde{P}_l^{OL}(t, T_l^{OL})$, and

- the residual equity position with a euro nominal of $N^{EQ}(t_0) \in \mathbb{R}_0^+$, which does not receive any dividend payments.

There is only one type of Pfandbriefe outstanding and, consequently, only one type of cover pool. Furthermore, all assets and liabilities are denominated in euros.

According to the above assumption, the bank has issued only one type of Pfandbriefe, the Pfandbrief type under consideration. In practice, Pfandbrief banks often have different types of Pfandbriefe outstanding. To avoid further model complexity, we account for such additional Pfandbrief business by adjusting funding and liquidation haircuts in a suitable manner, see Section 5.1.3 below. As long as the size of the additional Pfandbrief business is small, this is a viable assumption. For banks with considerable volumes of other Pfandbrief types, an explicit modelling of these positions might be necessary. A relaxation of this assumption is, however, beyond the scope of this work and left for future research. The bank's risky assets and its liabilities are modelled as positions in risky zero coupon bonds, an assumption which is not necessarily in line with payment profiles observed in practice.[4] Nevertheless, the zero coupon bond assumption is often made for modelling purposes, see for example Merton [124], Sünderhauf [148] or Liang et al [114], and avoids additional complexities arising from coupon payments. While for risky zero coupon bonds the assumption that default can only occur at maturity is still a reasonable approximation, it is less realistic in the case of risky coupon bonds where default may also be triggered by the inability to make coupon payments. Allowing for defaults prior to maturity would therefore result in increased model complexity with detailed information on the timing and size of coupon payments, especially for the bank's assets, often not being available in practice. The zero coupon bond assumption is therefore a reasonable approximation for our purpose. As the bank's assets and liabilities are all denominated in euros, Assumption 1 also implies that there is no currency risk. This assumption is made to avoid unnecessary additional model complexity[5] and is defensible as currency risk in German Pfandbrief programs is limited to a considerable extent, see Spangler and Werner [145], p. 42. For reasons of simplicity it is also assumed that there are no dividend payments to shareholders, but our model could be easily extended accordingly, under additional assumptions regarding the timing and size of these cash flows.

All in all, our balance sheet is a modified generalization of the balance sheet in Sünderhauf's [148] setup with only one Pfandbrief type, cf. Section 3.3.2.1. It allows for a sufficiently adequate representation of a Pfandbrief bank's typical risk profile and considers not only the assets' present values but also their cash flows, which correspond to the nominal repayments/recovery payments at maturity. It therefore fulfils all modelling requirements in Table 2.2.

To allow for a meaningful specification of our model, we need to make additional assumptions regarding the maturities of the bank's assets and liabilities.

---

[4]Jumbo Pfandbriefe typically have annual fix rate coupons paid in arrears and hard bullet redemptions and the vast majority of outstanding Pfandbriefe has either fixed or variable coupons, cf. vdp [163]. Furthermore, German mortgage assets often pay fixed coupons with fixed interest rate periods of 10 years, see Rudolf and Saunders [139], and may exhibit features such as amortizing nominals.

[5]A relaxation of the assumption would, amongst others, require an extension of the market environment in Section 4.3 to include foreign exchange rates and foreign currency interest rates.

Assumption 2 (Maturities of assets and liabilities)
For the bank's assets and liabilities it holds that

$$(i) \ T^{PB}_{n_{PB}} \leq T^{CPS}_{n_{CPS}}, \quad (ii) \ T^{OL}_{n_{OL}} \leq T^{OA}_{n_{OA}}, \quad (iii) \ T^{PB}_{n_{PB}} = T^{OL}_{n_{OL}}.$$

As that the bank's risky assets can only default at maturity, Assumption 2 guarantees that, at any time $t \in \mathcal{T}_d$, $t < T^{PB}_{n_{PB}} = T^{OL}_{n_{OL}}$, there is always at least one strategic cover pool asset and one other asset outstanding. It also prevents situations where Pfandbriefe are outstanding after all other liabilities have matured.

In the following, we assume that the simulation stops once the last Pfandbrief is due. As our focus is on Pfandbrief modelling, there is no need to consider time horizons beyond the maturity of the longest Pfandbrief.

Assumption 3 (Planned balance sheet liquidation)
The simulation stops at time $T_{\max} := T^{PB}_{n_{PB}}$ *and all remaining assets on the balance sheet are liquidated to repay the remaining debt, i.e. our simulation horizon is given by*

$$\mathcal{T} := \{t_0, ..., T_{\max}\} . \tag{4.2}$$

Asset liquidations in $T_{\max}$ (*planned asset liquidations*), which take place irrespective of whether or not an event of default occurs at that time, are to be distinguished from asset liquidations which are triggered by an event of bank or cover pool default at some time prior to $T_{\max}$ (*forced asset liquidations*). For more details, see Section 4.5 below.

The composition of the bank's balance sheet changes over time due to asset and liability maturities, stochastic market movements, asset liability management and potential default events. As a consequence, the time-t nominals, $t > t_0$, are all stochastic.

Time-t nominals. The stochastic time-t nominals, $t \in \mathcal{T}$, $t > t_0$, of the bank's assets and liabilities, which comprise the nominal repayments due at time t, are denoted by $N^{CPS}_i(t)$, $N^{CPL}(t)$, $N^{OA}_j(t)$, $N^{PB}_k(t)$ and $N^{OL}_l(t)$ for $i = 1, ..., n_{CPS}$, $j = 1, ..., n_{OA}$, $k = 1, ..., n_{PB}$ and $l = 1, ..., n_{OL}$. For $x \in \{CPS, OA, PB, OL\}$, we further set $N^x(t) := \sum_{i=1}^{n_x} N^x_i(t)$. The time-$t$ nominals are all $\mathcal{F}_{t-\Delta}$-measurable.

For modelling purposes, we make an additional assumption which affects the bank's balance sheet over time.

Assumption 4 (Run-off/no new business)
The bank does not engage in new business activities and only takes actions which are necessary in the context of maintenance of cover requirements, debt repayments and the reinvestment of excess cash inflows. In particular, no new Pfandbriefe or other liabilities are issued at any time $t \in \mathcal{T}$, assets are only purchased in the context of maintenance of overcollateralization and when excess cash inflows are reinvested, and asset sales only take place in case of a bank or cover pool default event or a planned balance sheet liquidation in $T_{\max}$. Apart from these cases, no assets are purchased or sold.

Allowing for new business would require additional assumptions regarding the nature of this business (e.g. volume and type) and its funding, which in turn has an impact on the bank's future balance sheet composition and therefore on model results. Depending on the related assumptions, the modelling of new business might also successively increase the simulation horizon as it involves the purchase of new assets and the issuance of additional liabilities. Therefore, we deem it more appropriate to consider a run-off situation which leaves the bank's risk profile "as it is", without a new business bias. Given the run-off assumption, the issuance of additional debt only makes sense in case of temporary liquidity shortages and the need to finance debt payments or to reestablish cover requirements. As we will see in Section 4.4 below, in our model such funding activities do not involve the issuance of additional Pfandbriefe or other liabilities. A collateralized liquidity line is drawn instead.

For the sake of simplicity, costs and taxes are ignored in the following. This assumption could, however, be easily relaxed by including the corresponding positions in our model, under additional assumptions regarding their timing and size.

Assumption 5 (No costs and taxes)
There are no taxes, issuer operating and cover pool administration costs.

4.3 Market Environment

In this section we specify the market environment through the stochastic dynamics of the risk-free interest rates (Section 4.3.1) and the creditworthiness of risky assets (Section 4.3.2) and derive the impact of these risk drivers on the present values and cash flows of the bank's risky assets (Section 4.3.3). We do so by ensuring that the required modelling features for the market environment as specified in Table 2.2 are fulfilled. While all relevant events in our simulation model happen at the discrete time steps in $\mathcal{T}$ as defined in Equation (4.2), our market environment is specified in a continuous time setting

$$\mathcal{T}_c := \left[t_0, T_{\max}^A \right], \quad \mathcal{T} \subset \mathcal{T}_c,$$

with $T_{\max}^A := \max(T_{n_{CPS}}^{CPS}, T_{n_{OA}}^{OA})$, which comprises the maturities of the bank's risky assets.

4.3.1 Risk-Free Interest Rates

In order to describe the evolution of the risk-free spot rate $r(t, T)$ over time, an interest rate model has to be chosen. The literature distinguishes *short rate models* such as the Vasicek model, the Hull-White model, the Black-Karasinksi model or the CIR model, from *market models* like the LIBOR market model and the swap market model. In the case of short rate models, the evolution of the whole yield curve solely depends on the short rate which is modelled by a one- or a multi-dimensional diffusion process. Market models, on the other hand, directly model rates observable in the market (LIBOR or swap rates). Market models are the natural choice for the pricing and hedging of complex interest rate derivatives. Nevertheless, short rate models are still used for many applications. For more details regarding interest rate models and their properties, we refer to Brigo and Mercurio [24] and Zagst [168].

In the end, the choice of the most suitable interest rate model depends on the intended purpose of use. In the following, we focus on the well-known short rate approach by Vasicek [154], which is very popular due to its analytical tractability and the availability of analytical valuation formulas for many instruments. Even though this model is not able to exactly reproduce a given initial term structure and cannot be fitted well to option prices, it remains our method of choice due to its analytical tractability. As our focus is not on pricing complex interest rate sensitive products but on Pfandbrief modelling, which involves the quantification of risks arising from asset liability mismatches (modelled as zero coupon bonds) under different interest rate scenarios, the Vasicek model is clearly sufficient. We therefore introduce the following notation.

Short rate. The short rate $r(t)$ at time t is the constant interest rate earned over an infinitesimally small period of time,

$$r(t) := -\lim_{\Delta_t \to 0} \frac{\ln P(t, t + \Delta_t)}{\Delta_t} = -\frac{\partial}{\partial T} \ln P(t, T)\big|_{T=t}, \quad \forall t \in \mathcal{T}_c,$$

given that all derivatives exist.

Vasicek model. In the Vasicek model, the short rate $r(t)$ evolves as an Ornstein-Uhlenbeck process with constant coefficients under the real-world measure $\mathbb{P}$:

$$dr(t) = \kappa_r \left(\theta_r - r(t)\right) dt + \sigma_r dW_r(t), \quad r(t_0) = r_0, \quad \forall t \in \mathcal{T}_c, \quad (4.3)$$

where κ_r, θ_r and σ_r are positive constants, $r_0 \in \mathbb{R}$ and $(W_r(t))_{t \in \mathcal{T}_c}$ a standard Brownian motion under $\mathbb{P}$.

The process in Equation (4.3) exhibits a positive drift for $r(t) < \theta_r$ and a negative drift for $r(t) > \theta_r$. This pushes $r(t)$ back to the level θ_r, which can be interpreted as the long term average of the short rate. The parameter κ_r measures the speed at which the process reverts to its long term average: the higher κ_r, the faster the mean reversion. By assuming that κ_r, θ_r and σ_r are positive constants, any potential time dependence or stochastic nature of these parameters is neglected.

In the Vasicek model, the future distribution of the short rate is known and its mean and variance can be determined analytically.

Proposition 4.1 (Short rate distribution in the Vasicek model)
From the short rate dynamics in Equation (4.3) it follows that, for all $s, t \in \mathcal{T}_c$, $s \le t$,

$$r(t) = r(s)e^{-\kappa_r(t-s)} + \theta_r \left(1 - e^{-\kappa_r(t-s)}\right) + \sigma_r \int_s^t e^{-\kappa_r(t-u)} dW(u),$$

and $r(t)$ conditional on $\mathcal{F}_s$ is normally distributed

$$r(t)|\mathcal{F}_s \sim \mathcal{N}\left(\mu_{r(t)|\mathcal{F}_s}, \sigma^2_{r(t)|\mathcal{F}_s}\right),$$

with

$$\mu_{r(t)|\mathcal{F}_s} := r(s)e^{-\kappa_r(t-s)} + \theta_r\left(1 - e^{-\kappa_r(t-s)}\right),$$

$$\sigma^2_{r(t)|\mathcal{F}_s} := \frac{\sigma_r^2}{2\kappa_r}\left(1 - e^{-2\kappa_r(t-s)}\right).$$

Proof. See Brigo and Mercurio [24], pp. 58ff.   □

Even though we are working in a finite time setting note that, for $t \to \infty$, we get $\lim_{t\to\infty} \mu_{r(t)|\mathcal{F}_s} = \theta_r$. This is why θ_r is referred to as the long term average. Similarly, the long term volatility is $\lim_{t\to\infty} \sigma^2_{r(t)|\mathcal{F}_s} = \frac{\sigma_r^2}{2\kappa_r}$. Proposition 4.1 also implies that there is a positive probability of negative interest rates, the magnitude of which depends on the actual choice of the parameters r_0, κ_r, θ_r and σ_r. The occurrence of negative interest rates often used to be argued to be the major drawback of the Vasicek model. Given recent market developments, this view has changed and negative interest rates are not considered as unrealistic any more.

Assumption 6 (Risk-free short rate dynamics under $\mathbb{P}$)
Under the real-world measure $\mathbb{P}$, the tradeable risk-free short rate evolves according to the Vasicek dynamics, i.e.

$$dr(t) = \kappa_r^P\left(\theta_r^P - r(t)\right)dt + \sigma_r dW_r^P(t), \quad r(t_0) = r_0, \quad \forall t \in \mathcal{T}_c, \qquad (4.4)$$

with κ_r^P, θ_r^P and σ_r being positive constants, $r_0 \in \mathbb{R}$ and $\left(W_r^P(t)\right)_{t\in\mathcal{T}_c}$ a standard Brownian motion under $\mathbb{P}$. Furthermore, the Radon-Nikodym derivative of the risk-neutral measure $\mathbb{Q}$ with respect to $\mathbb{P}$, restricted to $\mathcal{F}_t$, is given by

$$\left.\frac{d\mathbb{Q}}{d\mathbb{P}}\right|_{\mathcal{F}_t} = \exp\left(-\frac{1}{2}\int_0^t \gamma_r^2(s)ds - \int_0^t \gamma_r(s)dW_r^P(s)\right),$$

where $\gamma_r(t) := \lambda_1 r(t) + \lambda_2$ with

$$\lambda_1 := \frac{\kappa_r^Q - \kappa_r^P}{\sigma_r}, \quad \lambda_2 := \frac{\kappa_r^P \theta_r^P - \kappa_r^Q \theta_r^Q}{\sigma_r},$$

and κ_r^Q and θ_r^Q being positive constants.

It can be shown, see Proposition 4.2 below, that under the above assumptions the resulting dynamics under $\mathbb{Q}$ are the same as under $\mathbb{P}$, the only difference being the parameters θ_r and κ_r. For $\lambda_1 = \lambda_2 = 0$ the dynamics coincide.

Proposition 4.2 (Risk-free short rate dynamics under $\mathbb{Q}$)
Given Assumption 6, the dynamics of the risk-free short rate under the risk-neutral measure $\mathbb{Q}$ are given by

$$dr(t) = \kappa_r^Q \left(\theta_r^Q - r(t) \right) dt + \sigma_r dW_r^Q(t), \quad r(t_0) = r_0, \quad \forall t \in \mathcal{T}_c, \quad (4.5)$$

with r_0, σ_r, κ_r^Q and θ_r^Q as in Assumption 6 and $\left(W_r^Q(t) \right)_{t \in \mathcal{T}_c}$ being a standard Brownian motion under $\mathbb{Q}$. Furthermore, it holds that

$$\kappa_r^Q = \lambda_1 \sigma_r + \kappa_r^P \quad \text{and} \quad \theta_r^Q = \frac{\kappa_r^P \theta_r^P - \sigma_r \lambda_2}{\kappa_r^Q}.$$

Proof. From Assumption 6 and Girsanov's theorem, cf. Zagst [168], pp. 33–35, it follows that the stochastic process

$$dW_r^Q(t) = \gamma_r(t) dt + dW_r^P(t)$$

is a standard Brownian motion under $\mathbb{Q}$. Substituting $dW_r^P(t)$ in Equation (4.4) we get

$$\begin{aligned} dr(t) &= \kappa_r^P \left(\theta_r^P - r(t) \right) dt + \sigma_r \left(dW_r^Q(t) - \gamma_r(t) dt \right) \\ &= \kappa_r^P \theta_r^P dt - \kappa_r^P r(t) dt + \left(\kappa_r^P - \kappa_r^Q \right) r(t) dt + \left(\kappa_r^Q \theta_r^Q - \kappa_r^P \theta_r^P \right) dt \\ &\quad + \sigma_r dW_r^Q(t) \\ &= \kappa_r^Q \left(\theta_r^Q - r(t) \right) dt + \sigma_r dW_r^Q(t). \end{aligned}$$

Solving $\lambda_1 := \frac{\kappa_r^Q - \kappa_r^P}{\sigma_r}$ for κ_r^Q yields $\kappa_r^Q = \lambda_1 \sigma_r + \kappa_r^P$ and from $\lambda_2 := \frac{\kappa_r^P \theta_r^P - \kappa_r^Q \theta_r^Q}{\sigma_r}$ it follows that $\theta_r^Q = \frac{\kappa_r^P \theta_r^P - \sigma_r \lambda_2}{\kappa_r^Q}$. $\qquad \square$

The present values of risk-free zero coupon bonds can be calculated analytically in the Vasicek model.

Proposition 4.3 (Risk-free zero coupon bond price)
In the Vasicek model, the time-t present value of a risk-free zero coupon bond with maturity T is given by

$$P(t,T) = A(t,T)e^{-B(t,T)\cdot r(t)}, \quad t, T \in \mathcal{T}_c, \quad t \leq T, \tag{4.6}$$

with

$$A(t,T) := \exp\left\{ \left(\theta_r^Q - \frac{\sigma_r^2}{2(\kappa_r^Q)^2} \right) (B(t,T) - T + t) - \frac{\sigma_r^2}{4\kappa_r^Q} B(t,T)^2 \right\},$$

$$B(t,T) := \frac{1}{\kappa_r^Q} \left(1 - e^{-\kappa_r^Q(T-t)} \right).$$

Proof. See Brigo and Mercurio [24], p. 59.   □

Proposition 4.3 implies that the time-t present value of a risk-free zero coupon bond with maturity T only depends on the then-prevailing short rate level $r(t)$, the bond's remaining time to maturity $T - t$ and the risk-neutral short rate parameters κ_r^Q, θ_r^Q and σ_r. While the risk-neutral short rate parameters are constants and the bond's remaining time to maturity $T - t$ changes deterministically over time, the short rate level $r(t)$ is stochastic and follows the real-world dynamics as specified by Equation (4.4).

It can be shown, see Corollary 4.4 below, that the Vasicek model is a special case of the Hull-White extended Vasicek model as in Brigo and Mercurio [24], p. 73. This relationship will be needed later on.

Hull-White extended Vasicek model. In the Hull-White extended Vasicek model, the $\mathbb{Q}$-dynamics of the short rate $r(t)$ are given by

$$dr(t) = \left(\theta_r^H(t) - a_r^H r(t) \right) dt + \sigma_r^H dW_r^Q(t), \quad \forall t \in \mathcal{T}_c, \tag{4.7}$$

with $r(0) := r_0^H$ and a_r^H, σ_r^H being positive constants. The function $\theta_r^H(t)$ is chosen to exactly fit the term structure of interest rates being observed in the market,

$$\theta_r^H(t) := \frac{\partial f^M(0,t)}{\partial t} + a_r^H f^M(0,t) + \frac{\left(\sigma_r^H\right)^2}{2a_r^H}\left(1 - e^{-2a_r^H t}\right),$$

where $f^M(0,t) := -\frac{\partial \ln P^M(0,t)}{\partial t}$ is the market instantaneous forward rate at time 0 for maturity t and $P^M(0,t)$ the market discount factor for maturity t.

With respect to the Hull-White extended Vasicek model, the following corollary holds.

Corollary 4.4 (Hull-White extended Vasicek model)
If, under the risk-neutral measure $\mathbb{Q}$, the risk-free short rate evolves according to the Hull-White extended Vasicek dynamics in Equation (4.7) and the market discount factors $P^M(0,t)$ can be written as

$$P^M(0,T) := A(0,T)e^{-B(0,T)r_0}, \quad \forall t \in \mathcal{T}_c, \tag{4.8}$$

with $B(0,T)$ and $A(0,T)$ as in Proposition 4.3, choosing $a_r^H := \kappa_r^Q$, $\sigma_r^H := \sigma_r$ and $r_0^H := r_0$ yields the Vasicek dynamics as in Equation (4.5).

Proof. From Equation (4.8) it follows that

$$B(0,T) = \frac{1}{\kappa_r^Q}\left[1 - e^{-\kappa_r^Q T}\right],$$

$$\ln A(0,T) = \left(\theta_r^Q - \frac{\sigma_r^2}{2(\kappa_r^Q)^2}\right)(B(0,T) - T) - \frac{\sigma_r^2}{4\kappa_r^Q}B(0,T)^2$$

$$= \frac{\theta_r^Q}{\kappa_r^Q} - \frac{\theta_r^Q}{\kappa_r^Q}e^{-\kappa_r^Q T} - \theta_r^Q T - \frac{\sigma_r^2}{2(\kappa_r^Q)^3} + \frac{\sigma_r^2}{2(\kappa_r^Q)^3}e^{-\kappa_r^Q T}$$

$$+ \frac{\sigma_r^2}{2(\kappa_r^Q)^2}T - \frac{\sigma_r^2}{4(\kappa_r^Q)^3} + \frac{\sigma_r^2}{2(\kappa_r^Q)^3}e^{-\kappa_r^Q T} - \frac{\sigma_r^2}{4(\kappa_r^Q)^3}e^{-2\kappa_r^Q T},$$

$$\frac{\partial B(0,T)}{\partial T} = e^{-\kappa_r^Q T},$$

$$\frac{\partial \ln A(0,T)}{\partial T} = \theta_r^Q e^{-\kappa_r^Q T} - \theta_r^Q - \frac{\sigma_r^2}{(\kappa_r^Q)^2}e^{-\kappa_r^Q T} + \frac{\sigma_r^2}{2(\kappa_r^Q)^2} + \frac{\sigma_r^2}{2(\kappa_r^Q)^2}e^{-2\kappa_r^Q T}.$$

We therefore get

$$f^M(0,T) = -\frac{\partial \ln A(0,T)}{\partial T} + \frac{\partial B(0,T)}{\partial T}r_0$$

$$= -\theta_r^Q e^{-\kappa_r^Q T} + \theta_r^Q + \frac{\sigma_r^2}{(\kappa_r^Q)^2}e^{-\kappa_r^Q T} - \frac{\sigma_r^2}{2(\kappa_r^Q)^2} - \frac{\sigma_r^2}{2(\kappa_r^Q)^2}e^{-2\kappa_r^Q T}$$

$$+ r_0 e^{-\kappa_r^Q T},$$

$$\frac{\partial f^M(0,T)}{\partial T} = \theta_r^Q \kappa_r^Q e^{-\kappa_r^Q T} - \frac{\sigma_r^2}{\kappa_r^Q}e^{-\kappa_r^Q T} + \frac{\sigma_r^2}{\kappa_r^Q}e^{-2\kappa_r^Q T} - r_0 \kappa_r^Q e^{-\kappa_r^Q T}.$$

Setting $a_r^H = \kappa_r^Q$ and $\sigma_r^H = \sigma_r$, the function $\theta_r^H(t)$ reduces to

$$\theta_r^H(t) = \frac{\partial f^M(0,t)}{\partial t} + \kappa_r^Q f^M(0,t) + \frac{\sigma_r^2}{2\kappa_r^Q}\left(1 - e^{-2\kappa_r^Q t}\right)$$

$$= \theta_r^Q \kappa_r^Q e^{-\kappa_r^Q T} - \frac{\sigma_r^2}{\kappa_r^Q}e^{-\kappa_r^Q T} + \frac{\sigma_r^2}{\kappa_r^Q}e^{-2\kappa_r^Q T} - r_0 \kappa_r^Q e^{-\kappa_r^Q T} - \kappa_r^Q \theta_r^Q e^{-\kappa_r^Q T}$$

$$+ \kappa_r^Q \theta_r^Q + \frac{\sigma_r^2}{\kappa_r^Q}e^{-\kappa_r^Q T} - \frac{\sigma_r^2}{2\kappa_r^Q} - \frac{\sigma_r^2}{2\kappa_r^Q}e^{-2\kappa_r^Q T} + r_0 \kappa_r^Q e^{-\kappa_r^Q T} + \frac{\sigma_r^2}{2\kappa_r^Q}$$

$$- \frac{\sigma_r^2}{2\kappa_r^Q}e^{-2\kappa_r^Q t}$$

$$= \kappa_r^Q \theta_r^Q,$$

and with $r_0^H = r_0$, the dynamics in Equation (4.7) read

$$dr(t) = \kappa_r^Q\left(\theta_r^Q - r(t)\right)dt + \sigma_r dW_r^Q(t), \quad r(0) = r_0, \quad \forall t \in \mathcal{T}_c,$$

i.e. they are equal to the Vasicek dynamics in Equation (4.5). □

The Vasicek model is an endogenous term structure model, which means that the interest rate term structure is a model output rather than a model input.[6] The analytical expression for the model-implied term structure can be easily derived as follows.

---

[6]This is as opposed to other interest rate models (such as the Libor Market Model or the Hull-White extended Vasicek model) which are able to reproduce a given initial yield curve exactly.

Remark 4.5 (Risk-free spot rate)
In the Vasicek model, the risk-free spot rate prevailing at time t for the maturity T is given by

$$r(t,T) = -\frac{\ln P(t,T)}{T-t} = -\frac{\ln A(t,T)}{T-t} + \frac{B(t,T)}{T-t}r(t), \quad t,T \in \mathcal{T}_c, \quad t < T.$$
(4.9)

Under the assumptions made so far, the risk-free zero coupon bond price $P(t,T)$ is lognormally distributed with mean and variance depending both on the risk-neutral and the real-world Vasicek parameters.

Proposition 4.6 (Distribution of the risk-free zero coupon bond price)
Given Assumption 6, the time-t present value of a risk-free zero coupon bond with maturity T, conditional on $\mathcal{F}_s$, is lognormally distributed

$$\ln P(t,T)|\mathcal{F}_s \sim \mathcal{N}\left(\tilde{\mu}_{P(t,T)|\mathcal{F}_s}, \tilde{\sigma}^2_{P(t,T)|\mathcal{F}_s}\right),$$

for $s,t,T \in \mathcal{T}_c$, $s \leq t \leq T$, with

$$\begin{aligned}
\tilde{\mu}_{P(t,T)|\mathcal{F}_s} &:= \ln A(t,T) - B(t,T)\mu_{r(t)|\mathcal{F}_s}, \\
\tilde{\sigma}^2_{P(t,T)|\mathcal{F}_s} &:= \sigma^2_{r(t)|\mathcal{F}_s}B(t,T)^2,
\end{aligned}$$
(4.10)

and the mean and the variance of $P(t,T)$ are given by

$$\mu_{P(t,T)|\mathcal{F}_s} := \exp\left(\tilde{\mu}_{P(t,T)|\mathcal{F}_s} + \frac{1}{2}\tilde{\sigma}^2_{P(t,T)|\mathcal{F}_s}\right),$$

$$\sigma^2_{P(t,T)|\mathcal{F}_s} := \exp\left(2\tilde{\mu}_{P(t,T)|\mathcal{F}_s} + \tilde{\sigma}^2_{P(t,T)|\mathcal{F}_s}\right) \cdot \left(\exp\left(\tilde{\sigma}^2_{P(t,T)|\mathcal{F}_s}\right) - 1\right).$$
(4.11)

Proof. According to Assumption 6, the real-world dynamics of the short rate are given by Equation (4.4) and from Proposition 4.1 and 4.3 we know that $r(t)|\mathcal{F}_s \sim \mathcal{N}\left(\mu_{r(t)|\mathcal{F}_0}, \sigma^2_{r(t)|\mathcal{F}_s}\right)$ and $\ln P(t,T) = \ln A(t,T) - B(t,T) \cdot r(t)$. For $X \sim \mathcal{N}(\mu,\sigma)$, the transformed random variable $Y := aX + b$ with $a \neq 0$ and $b \in \mathbb{R}$ is also normally distributed with $Y \sim \mathcal{N}\left(a\mu + b, a^2\sigma^2\right)$ and it follows that $\ln P(t,T)|\mathcal{F}_s \sim \mathcal{N}\left(\tilde{\mu}_{P(t,T)|\mathcal{F}_s}, \tilde{\sigma}^2_{P(t,T)|\mathcal{F}_s}\right)$ with $\tilde{\mu}_{P(t,T)|\mathcal{F}_s}$ and

$\tilde{\sigma}^2_{P(t,T)|\mathcal{F}_s}$ as in (4.10). For a lognormal random variable X, $\ln X \sim \mathcal{N}\left(\mu, \sigma^2\right)$, it holds that $\mathbb{E}[X] = e^{\mu + \frac{\sigma^2}{2}}$ and $\mathbb{V}[X] = e^{2\mu + \sigma^2}(e^{\sigma^2} - 1)$. The mean and the variance of $P(t,T)$ are therefore given by $\mu_{P(t,T)|\mathcal{F}_s}$ and $\tilde{\sigma}^2_{P(t,T)|\mathcal{F}_s}$ as in (4.11). □

The implications of Proposition 4.6 for the risk-free spot rate are straightforward.

Remark 4.7 (Distribution of the risk-free spot rate)
Given Assumption 6, the risk-free spot rate prevailing at time t for the maturity T, conditional on $\mathcal{F}_s$, is normally distributed

$$r(t,T)|\mathcal{F}_s \sim \mathcal{N}\left(\tilde{\mu}_{r(t,T)|\mathcal{F}_s}, \tilde{\sigma}^2_{r(t,T)|\mathcal{F}_s}\right),$$

for $s, t, T \in \mathcal{T}_c$, $s \leq t \leq T$, with

$$\tilde{\mu}_{r(t,T)|\mathcal{F}_s} := -\frac{1}{T-t}\tilde{\mu}_{P(t,T)|\mathcal{F}_s},$$

$$\tilde{\sigma}^2_{r(t,T)|\mathcal{F}_s} := \frac{1}{(T-t)^2}\tilde{\sigma}^2_{P(t,T)|\mathcal{F}_s}.$$

Proof. The statement follows directly from $r(t,T) = -\frac{1}{T-t}\ln P(t,T)$. □

In the Vasicek model, the parameters are specified to be positive constants. Allowing for $\sigma_r = 0$ corresponds to the special case of deterministic short rate dynamics with deterministic zero coupon bond prices and spot rates. The additional requirement $\theta_r^P = \theta_r^Q = r_0$ results in constant interest rates.

Remark 4.8 (Special cases of the Vasicek model)
Setting $\sigma_r = 0$ in Equation (4.4), the real-world short rate dynamics become deterministic, i.e. for $s, t \in \mathcal{T}_c$, $s \leq t$,

$$r(t) = r(s)e^{-\kappa_r^P(t-s)} + \theta_r^P(1 - e^{-\kappa_r^P(t-s)}),$$

and with $\kappa_r^P = \kappa_r^Q$ and $\theta_r^P = \theta_r^Q$ the zero coupon bond price and the risk-free spot rate simplify to

$$P(t,T) = \mathbb{E}^{\mathbb{Q}} \left[e^{-\int_t^T r(\tau)d\tau} \middle| \mathcal{F}_t \right]$$

$$= \exp \left\{ -r(t) \int_t^T e^{-\kappa_r^P(\tau-t)} d\tau - \theta_r^P \int_t^T (1 - e^{-\kappa_r^P(\tau-t)}) d\tau \right\}$$

$$= \exp \left\{ -\frac{r(t)}{\kappa_r^P} \left(1 - e^{-\kappa_r^P(T-t)} \right) - \theta_r^P (T - t) + \frac{\theta_r^P}{\kappa_r^P} \left(1 - e^{-\kappa_r^P(T-t)} \right) \right\}$$

$$= \exp \left\{ -B(t,T) \cdot r(t) + \theta_r^Q (B(t,T) - T + t) \right\},$$

$$r(t,T) = -\frac{\ln P(t,T)}{T-t} = -\frac{1}{T-t} \left(-B(t,T) \cdot r(t) + \theta_r^Q (B(t,T) - T + t) \right).$$

The additional requirement $\theta_r^P = \theta_r^Q = r_0$ results in a constant short rate $r(t) = r_0$, $P(t,T) = e^{-r_0 \cdot (T-t)}$ and $r(t,T) = r_0$.

4.3.2 Asset Creditworthiness

To model the creditworthiness of the bank's risky assets, an asset credit risk model has to be chosen. Potential alternatives include the model by Jarrow et al [96], which takes into account credit migrations with deterministic credit spreads, or the one by Dubrana [50], which includes credit migrations with stochastic credit spreads. In our Pfandbrief model, we follow Sünderhauf [148] and assume that the creditworthiness of a risky zero coupon bond is driven by a state variable process, but we ignore the jump component and the stochastic volatility used in his model and restrict ourselves to geometric Brownian motions. As a consequence, we obtain analytical solutions for asset present values under stochastic interest rates and creditworthiness and do not require additional Monte Carlo simulations for pricing purposes. We also avoid additional uncertainties arising from parameter estimation in jump processes and obtain explicit formulas for asset default probabilities and losses given default, which facilitates the calibration of the state variable process to given default parameters.

Assumption 7 (Creditworthiness of a risky zero coupon bond)
For a risky zero coupon bond, the following holds:

(a) *The payoff of the risky zero coupon bond at its maturity $T \in \mathcal{T}_c$ is given by $\tilde{P}(T,T) := \min(1; Z(T))$, where $Z(T)$ is the time-T-value of an associated tradeable state variable, i.e. some firm or real estate value. The risky zero coupon bond cannot default prior to maturity and at time T it defaults if and only if $Z(T) < 1$.*

(b) Under the real-world measure $\mathbb{P}$, the dynamics of the state variable process are given by

$$dZ(t) = \mu_Z Z(t)dt + \sigma_Z Z(t)dW_Z^P(t), \quad Z(0) = Z_0, \quad \forall t \in \mathcal{T}_c,$$

with Z_0 and σ_Z being positive constants, $\mu_Z \in \mathbb{R}$ and $\left(W_Z^P(t)\right)_{t \in \mathcal{T}_c}$ a standard Brownian motion under $\mathbb{P}$.

(c) The $\mathbb{P}$-dynamics of the state variable Z and the risk-free short rate r are correlated according to $dW_r^P(t)dW_Z^P(t) = \rho dt$ with $\rho \in (-1, 1)$.

As the state variable follows a geometric Brownian motion, its distribution at the zero coupon bond's maturity is known.

Remark 4.9 (Distribution of the state variable)
At the risky zero coupon bond's maturity $T \in \mathcal{T}_c$, the value of the state variable Z is lognormally distributed

$$\ln Z(T)|\mathcal{F}_0 \sim \mathcal{N}\left(\tilde{\mu}_{Z(T)}, \tilde{\sigma}_{Z(T)}^2\right),$$

with

$$\tilde{\mu}_{Z(T)} := \ln Z_0 + \left(\mu_Z - 0.5 \cdot \sigma_Z^2\right)T,$$
$$\tilde{\sigma}_{Z(T)}^2 := \sigma_Z^2 T,$$

and the mean and the variance of $Z(T)$ are given by[7]

$$\mu_{Z(T)} := \exp\left(\tilde{\mu}_{Z(t)} + 0.5 \cdot \tilde{\sigma}_{Z(T)}^2\right),$$
$$\sigma_{Z(T)}^2 := \exp\left(2\tilde{\mu}_{Z(T)} + \tilde{\sigma}_{Z(T)}^2\right) \cdot \left(\exp\left(\tilde{\sigma}_{Z(T)}^2\right) - 1\right).$$

The zero coupon bond's default parameters which are the basis for the calibration of the state variable processes, cf. Section 5.2.2 below, can now be determined.

---

[7]This follows from the fact that, for a lognormal random variable X, $\ln X \sim \mathcal{N}\left(\mu, \sigma^2\right)$, it holds that $\mathbb{E}[X] = e^{\mu + \frac{\sigma^2}{2}}$ and $\mathbb{V}[X] = e^{2\mu + \sigma^2}(e^{\sigma^2} - 1)$.

Lemma 4.10 (Default parameters of a risky zero coupon bond)
The lifetime default probability and the lifetime loss given default of a risky zero coupon bond with maturity $T \in \mathcal{T}_c$ and associated state variable Z are given by

$$\pi(T) = \Phi\left(-\frac{\tilde{\mu}_{Z(T)}}{\tilde{\sigma}_{Z(T)}}\right),$$

$$L(T) = 1 - \frac{\mu_{Z(T)}}{\pi(T)} \cdot \Phi(\Phi^{-1}(\pi(T)) - \tilde{\sigma}_{Z(T)}),$$

if $T > t_0$, with Φ being the cumulative distribution function of the standard normal distribution. In the special case where $T = t_0$ it holds that

$$\pi(T) = \begin{cases} 1, & \text{if } Z_0 e^{\mu_Z T} < 1, \\ 0, & \text{otherwise.} \end{cases}$$

$$L(T) = \begin{cases} 1 - Z_0 e^{\mu_Z T}, & \text{if } Z_0 e^{\mu_Z T} < 1, \\ 0, & \text{otherwise.} \end{cases}$$

Proof. The risky zero coupon bond defaults if and only if $Z(T) < 1$, i.e. its default probability and loss given default are given by

$$\pi(T) = \mathbb{P}\left[Z(T) < 1\right] \quad \text{and} \quad L(T) = 1 - \mathbb{E}^{\mathbb{P}}\left[Z(T)\big|Z(T) < 1\right].$$

For $T > t_0$ it follows from Remark 4.9 that

$$\pi(T) = \mathbb{P}\left[\ln Z(T) < 0\right] = \Phi\left(-\frac{\tilde{\mu}_{Z(T)}}{\tilde{\sigma}_{Z(T)}}\right),$$

and it holds that

$$L(T) = 1 - \frac{\mathbb{E}^{\mathbb{P}}\left[Z(T) \cdot \mathbb{I}_{\{Z(T)<1\}}\right]}{\mathbb{P}\left[Z(T) < 1\right]} = 1 - \frac{\mathbb{E}^{\mathbb{P}}\left[Z(T)\right]}{\pi(T)} + \frac{\mathbb{E}^{\mathbb{P}}\left[Z(T) \cdot \mathbb{I}_{\{Z(T)\geq 1\}}\right]}{\pi(T)},$$

with $\mathbb{E}^{\mathbb{P}}\left[Z(T)\right] = \mu_{Z(T)}$. Using $\Phi(x) = 1 - \Phi(-x) \ \forall x \in \mathbb{R}$ we get

$$\mathbb{E}^{\mathbb{P}}\left[Z(T)\cdot \mathbb{I}_{\{Z(T)\geq 1\}}\right] = \mathbb{E}^{\mathbb{P}}\left[(Z(T)-1)^{+} + 1\cdot \mathbb{I}_{\{Z(t)\geq 1\}}\right]$$

$$= \mathbb{E}^{\mathbb{P}}\left[(Z(T)-1)^{+}\right] + \mathbb{P}\left[Z(T)\geq 1\right]$$

$$= \mathbb{E}^{\mathbb{P}}\left[(Z(T)-1)^{+}\right] + \Phi\left(\frac{\tilde{\mu}_{Z(T)}}{\tilde{\sigma}_{Z(T)}}\right),$$

where $X^{+} := \mathbb{I}_{\{X\geq 0\}}\cdot X$. As $Z(T)|\mathcal{F}_0$ is lognormally distributed, it follows that[8]

$$\mathbb{E}^{\mathbb{P}}\left[Z(T)\cdot \mathbb{I}_{\{Z(T)\geq 1\}}\right] = e^{\tilde{\mu}_{Z(T)}+\frac{1}{2}\tilde{\sigma}_{Z(T)}^{2}}\cdot \Phi\left(\frac{\tilde{\mu}_{Z(T)}}{\tilde{\sigma}_{Z(T)}}+\tilde{\sigma}_{Z(T)}\right)$$

$$= \mu_{Z(T)}\cdot \Phi\left(-\Phi^{-1}(\pi(T))+\tilde{\sigma}_{Z(T)}\right).$$

All in all, we have

$$L(T) = 1 - \frac{\mu_{Z(T)}}{\pi(T)}\cdot \left(1 - \Phi\left(-\Phi^{-1}(\pi(T))+\tilde{\sigma}_{Z(T)}\right)\right)$$

and, again with $\Phi(-x) = 1 - \Phi(x) \ \forall x \in \mathbb{R}$, this results in

$$L(T) = 1 - \frac{\mu_{Z(T)}}{\pi(T)}\cdot \Phi(\Phi^{-1}(\pi(T))-\tilde{\sigma}_{Z(T)}).$$

In the special cases where $T = t_0$ we obtain $\ln Z(T)|\mathcal{F}_0 \sim \mathcal{N}\left(\tilde{\mu}_{Z(T)},0\right)$, i.e. $Z(T)$ is deterministic and $Z(T) = \mathbb{E}^{\mathbb{P}}\left[Z(T)\right] = Z_0 e^{\mu_Z T} =: \mu_{Z(T)}^{d}$. Then, by definition, $\pi(T) = 1$ if $\mu_{Z(T)}^{d} < 1$ and $\pi(T) = 0$ otherwise. In the former case, $L(T) = 1 - \mu_{Z(T)}^{d}$ and in the latter case $L(T) = 0$. □

We are not only interested in the final payoff of the bank's risky zero coupon bond at its maturity T but also in its present value prior to maturity. Therefore, we make the following assumption:

---

[8]For a lognormally distributed random variable X and some constant $\tilde{K} > 0$, it holds that

$$\mathbb{E}\left[\left(X-\tilde{K}\right)^{+}\right] = e^{m+\frac{1}{2}v^2}\Phi\left(\frac{m-\ln(\tilde{K})+v^2}{v}\right) - \tilde{K}\phi\left(\frac{m-\ln\tilde{K}}{v}\right),$$

with $m := \mathbb{E}\left[\ln X\right]$ and $v := \mathbb{V}\left[\ln X\right]$ being the expectation and the variance of $\ln X$, cf. Brigo and Mercurio [24], p. 919 f.

Assumption 8 (Joint dynamics of the risk drivers under $\mathbb{P}$)
*Under the real-world measure $\mathbb{P}$, the joint dynamics of the short rate r and
the state variable Z are given by*

$$dY(t) = \mu_Y^P(t)dt + \Sigma_Y(t)dW_Y^P(t), \quad \forall t \in \mathcal{T}_c, \tag{4.12}$$

where $Y(t) := (r(t), Z(t))^{\top}$, $r(0) := r_0$, $Z(0) := Z_0$ and

$$\mu_Y^P(t) := \begin{pmatrix} \kappa_r^P \left(\theta_r^P - r(t) \right) \\ \mu_Z Z(t) \end{pmatrix},$$

$$\Sigma_Y(t) := \begin{pmatrix} \sigma_r & 0 \\ \sigma_Z \rho Z(t) & \sigma_Z \sqrt{1-\rho^2} Z(t) \end{pmatrix},$$

*with r_0, κ_r^P, θ_r^P, σ_r as in Assumption 6, Z_0, σ_Z, μ_Z, ρ as in Assumption
7, and $\left(W_Y^P(t) \right)_{t \in \mathcal{T}_c}$ a two-dimensional standard Brownian motion under $\mathbb{P}$.
The Radon-Nikodym derivative of the risk-neutral measure $\mathbb{Q}$ with respect to
$\mathbb{P}$, restricted to $\mathcal{F}_t$, is given by*

$$\left. \frac{d\mathbb{Q}}{d\mathbb{P}} \right|_{\mathcal{F}_t} = \exp\left(-\frac{1}{2} \int_0^t \|\gamma_Y(s)\|^2 ds - \int_0^t \gamma_Y^{\top}(s) dW_Y^P(s) \right),$$

where $\gamma_Y(s) := (\gamma_r(s), \gamma_Z(s))^{\top}$ with $\gamma_r(s)$ as in Assumption 6 and

$$\gamma_Z(s) := \frac{\mu_Z - r(s) - \sigma_Z \rho \gamma_r(s)}{\sigma_Z \sqrt{1-\rho^2}}.$$

Remark 4.11 (Joint dynamics of the risk drivers under $\mathbb{P}$)
*Under the real-world measure $\mathbb{P}$, the joint dynamics of the short rate r and
the state variable Z in (4.12) can be rewritten as*

$$dr(t) = \kappa_r^P \left(\theta_r^P - r(t) \right) dt + \sigma_r dW_r^P(t),$$

$$dZ(t) = \mu_Z Z(t)dt + \sigma_Z \rho Z(t)dW_r^P(t) + \sigma_Z \sqrt{1-\rho^2} Z(t)dW_{\perp}^P(t)$$

$$= \mu_Z Z(t)dt + \sigma_Z Z(t)dW_Z^P(t),$$

*with $dW_r^P(t)$ and $dW_{\perp}^P(t)$ being two independent standard Brownian motions
under $\mathbb{P}$ and $dW_Z^P(t) := \rho dW_r^P(t) + \sqrt{1-\rho^2}dW_{\perp}^P$. Assumption 8a therefore
subsumes the corresponding assumptions on the dynamics of $r(t)$ and $Z(t)$
as made in Assumptions 6 and 7.*

The resulting $\mathbb{Q}$-dynamics can be easily derived. The correlation between the processes r and Z is not affected by the change of measure.

Proposition 4.12 (Joint dynamics of the risk drivers under $\mathbb{Q}$)
The joint dynamics of the short rate r and the state variable Z under the risk-neutral measure $\mathbb{Q}$ are given by

$$dY(t) = \mu_Y^{\mathbb{Q}}(t)dt + \Sigma_Y(t)dW_Y^{\mathbb{Q}}(t), \quad \forall t \in \mathcal{T}_c, \tag{4.13}$$

with $Y(t)$, $\Sigma_Y(t)$ as in Assumption 8, $\mu_Y^{\mathbb{Q}}(t) := \left(\kappa_r^{\mathbb{Q}} \left(\theta_r^{\mathbb{Q}} - r(t) \right), r(t)Z(t) \right)^{\top}$, $\kappa_r^{\mathbb{Q}}$, $\theta_r^{\mathbb{Q}}$ as in Assumption 6 and $\left(W_Y^{\mathbb{Q}}(t) \right)_{t \in \mathcal{T}_c}$ a two-dimensional standard Brownian motion under $\mathbb{Q}$.

Proof. Given Assumption 8, it follows from Girsanov's theorem (cf. Zagst [168], pp. 33–35) that the stochastic process

$$dW_Y^{\mathbb{Q}}(t) = \gamma_Y(t)dt + dW_Y^{P}(t)$$

is a two-dimensional standard Brownian motion under $\mathbb{Q}$. Substituting $dW_Y^{P}(t)$ in Equation (4.12) we get

$$dY(t) = \mu_Y^{P}(t)dt - \Sigma_Y(t)\gamma_Y(t)dt + \Sigma_Y(t)dW_Y^{\mathbb{Q}}(t)$$

and therefore

$$\mu_Y^{P}(t) - \Sigma_Y(t)\gamma_Y(t) = \begin{pmatrix} \kappa_r^{P} \left(\theta_r^{P} - r(t) \right) - \sigma_r \gamma_r(t) \\ \mu_Z Z(t) - \sigma_Z \rho Z(t)\gamma_r(t) - \sigma_Z \sqrt{1 - \rho^2} Z(t)\gamma_Z(t) \end{pmatrix}$$
$$= \begin{pmatrix} \kappa_r^{\mathbb{Q}} \left(\theta_r^{\mathbb{Q}} - r(t) \right) \\ r(t)Z(t). \end{pmatrix} = \mu_Y^{\mathbb{Q}}(t).$$

$\square$

The present value of risky zero coupon bonds can now be determined as follows.

Lemma 4.13 (Risky zero coupon bond price)
At time $t < T$, $t, T \in \mathcal{T}_c$, the present value of a risky zero coupon bond with maturity T and associated state variable Z is given by

$$\tilde{P}(t,T) = P(t,T) - V^{\text{Put}}(t,T,1,Z(t),\sigma_Z,r(t),\theta_r^Q,\kappa_r^Q,\sigma_r,\rho), \qquad (4.14)$$

with

$$
\begin{aligned}
V^{\text{Put}}(t,T,K,Z,\sigma_Z,r,\theta_r^Q,\kappa_r^Q,\sigma_r,\rho) := \\
K \cdot P(t,T) \cdot \Phi(-d_2) - Z \cdot \Phi(-d_1),
\end{aligned}
\qquad (4.15)
$$

and

$$
\begin{aligned}
d_1 &:= \frac{1}{\tilde{v}(t,T)} \left(\ln\left(\frac{Z}{K \cdot P(t,T)} \right) + \frac{1}{2} \cdot \tilde{v}^2(t,T) \right), \\
d_2 &:= d_1 - \tilde{v}(t,T), \\
\tilde{v}^2(t,T) &:= \sigma_Z^2 (T-t) + 2w(t,T) + V(t,T), \\
w(t,T) &:= \rho \frac{\sigma_r \sigma_Z}{\kappa_r^Q} \left[T - t - \frac{1}{\kappa_r^Q}\left(1 - e^{-\kappa_r^Q(T-t)} \right) \right], \\
V(t,T) &:= \frac{\sigma_r^2}{(\kappa_r^Q)^2} \left[T - t + \frac{2}{\kappa_r^Q} \cdot e^{-\kappa_r^Q(T-t)} - \frac{1}{2\kappa_r^Q} e^{-2\kappa_r^Q(T-t)} - \frac{3}{2\kappa_r^Q} \right].
\end{aligned}
$$
$$(4.16)$$

Proof. Starting from $\tilde{P}(T,T) = \min(1, Z(T)) = 1 - (K - Z(T))^+$ with $K := 1$, the present value of the risky zero coupon bond can be written as

$$
\begin{aligned}
\tilde{P}(t,T) &= \mathbb{E}^Q \left[\tilde{P}(T,T) \cdot e^{-\int_t^T r(s)ds} \Big| \mathcal{F}_t \right] \\
&= \mathbb{E}^Q \left[e^{-\int_t^T r(s)ds} \Big| \mathcal{F}_t \right] - \mathbb{E}^Q \left[(K - Z(T))^+ \cdot e^{-\int_t^T r(s)ds} \Big| \mathcal{F}_t \right] \\
&= P(t,T) - U(t,T),
\end{aligned}
$$

where $P(t,T)$ is as in Equation (4.6) and

$$U(t,T) := \mathbb{E}^Q \left[(K - Z(T))^+ \cdot e^{-\int_t^T r(s)ds} \Big| \mathcal{F}_t \right].$$

As $(K - Z(T))^+$ is the payoff of a European put option on $Z(T)$ with strike

K, $U(t,T)$ corresponds to the option's present value. This means that it has to be shown that $U(t,T) \overset{!}{=} K \cdot P(t,T) \cdot \Phi(-d_2) - Z(t) \cdot \Phi(-d_1)$, with d_1 and d_2 as in (4.16). We do so in four steps:

Step 1: Changing from the risk-neutral measure $\mathbb{Q}$ to the forward measure $\mathbb{Q}^T$ we get, cf. Brigo and Mercurio [24], p. 30,

$$
U(t,T) = P(t,T) \cdot \mathbb{E}^{\mathbb{Q}^T} \left[(K - Z(T))^+ \Big| \mathcal{F}_t \right]
$$

$$
= Z(t) \cdot \mathbb{E}^{\mathbb{Q}^T} \left[\left(\frac{P(t,T) \cdot K}{Z(t)} - \frac{P(t,T) \cdot Z(T)}{Z(t)} \right)^+ \Big| \mathcal{F}_t \right]
$$

$$
= Z(t) \cdot \mathbb{E}^{\mathbb{Q}^T} \left[\left(\tilde{K}(t,T) - X(t,T) \right)^+ \Big| \mathcal{F}_t \right],
$$

with $\tilde{K}(t,T) := \frac{P(t,T) \cdot K}{Z(t)}$ and $X(t,T) := \frac{P(t,T) \cdot Z(T)}{Z(t)}$.

Step 2: As the risk-neutral Vasicek dynamics in Equation (4.5) are a special case of the ones in the Hull-White extended Vasicek model, see Corollary 4.4, we can use the results obtained by Brigo and Mercurio [24], pp. 886–887, which state that the following holds under the measure $\mathbb{Q}^T$:

$$
\ln \left(\frac{Z(T)}{Z(t)} \right) = \frac{1 - e^{-a_r^H (T-t)}}{a_r^H} x(t) + \underbrace{\frac{\sigma_r^H}{a_r^H} \int_t^T \left(1 - e^{-a_r^H (T-u)} \right) d\tilde{W}^T(u)}_{(*)}
$$

$$
- \frac{(\sigma_r^H)^2}{a_r^H} \int_t^T \int_t^u e^{-a_r^H (u-s)} \left(1 - e^{-a_r^H (T-s)} \right) ds \, du
$$

$$
+ \int_t^T f^M(0,u) du + \frac{(\sigma_r^H)^2}{2(a_r^H)^2} \int_t^T \left(1 - e^{-a_r^H u} \right)^2 du
$$

$$
- \rho \frac{\sigma_r^H \sigma_Z}{a_r^H} \int_t^T \left(1 - e^{-a_r^H (T-u)} \right) du - \frac{1}{2} \sigma_Z^2 (T-t)
$$

$$
+ \sigma_Z \sqrt{1 - \rho^2} \underbrace{\left(\tilde{Z}^T(T) - \tilde{Z}^T(t) \right)}_{(**)}
$$

$$
+ \sigma_Z \rho \underbrace{\left(\tilde{W}^T(T) - \tilde{W}^T(t) \right)}_{(***)}, \qquad \forall t < T,
$$

with $(\tilde{W}^T(t))_{t\in\mathcal{T}_c}$ and $(\tilde{Z}^T(t))_{t\in\mathcal{T}_c}$ being two independent Brownian motions under $\mathbb{Q}^T$ and $x(t)$ being $\mathcal{F}_t$-measurable. From the properties of Brownian motions it follows that, given $\mathcal{F}_t$, the expressions in (*), (**) and (***) are normally distributed and we get

$$\ln\left(\frac{Z(T)}{Z(t)}\right)\bigg|\mathcal{F}_t \sim \mathcal{N}\left(\bar{\mu}, \bar{\sigma}\right),$$

for some $\bar{\mu}, \bar{\sigma} \in \mathbb{R}$. With

$$\ln X(t,T) = \ln\left(\frac{Z(T)}{Z(t)}\right) + \ln P(t,T)$$

we conclude that $X(t,T)|\mathcal{F}_t$ is lognormally distributed under $\mathbb{Q}^T$.

Step 3: As $X(t,T)|\mathcal{F}_t$ is lognormally distributed under $\mathbb{Q}^T$, it follows that[9]

$$\mathbb{E}^{\mathbb{Q}^T}\left[\left(\tilde{K}(t,T) - X(t,T)\right)^+\bigg|\mathcal{F}_t\right]$$
$$= -e^{c(t,T)}\Phi\left(-\frac{f(t,T)}{b(t,T)}\right) + \tilde{K}(t,T)\Phi\left(-\frac{e(t,T)}{b(t,T)}\right),$$

with

$$c(t,T) := a(t,T) + \frac{1}{2}b(t,T)^2,$$
$$e(t,T) := a(t,T) - \ln\tilde{K}(t,T),$$
$$f(t,T) := a(t,T) - \ln\tilde{K}(t,T) + b(t,T)^2 = e(t,T) + b(t,T)^2,$$
$$a(t,T) := \mathbb{E}^{\mathbb{Q}^T}\left[\ln X(t,T)\right],$$
$$b(t,T)^2 := \mathbb{V}^{\mathbb{Q}^T}\left[\ln X(t,T)\right].$$

Step 4: Using again the fact that the risk-neutral Vasicek dynamics in Equation (4.5) are a special case of the ones in the Hull-White extended Vasicek model, we can refer to the formulas for the conditional expectation

---

[9] Again, we use the fact that, for a lognormally distributed random variable X and some constant $\tilde{K} > 0$, it holds that

$$\mathbb{E}\left[\left(\tilde{K} - X\right)^+\right] = -e^{m+\frac{1}{2}v^2}\Phi\left(-\frac{M - \ln(\tilde{K}) + v^2}{v}\right) + \tilde{K}\Phi\left(-\frac{M - \ln\tilde{K}}{v}\right),$$

with $m := \mathbb{E}\left[\ln X\right]$ and $v := \mathbb{V}\left[\ln X\right]$, cf. Brigo and Mercurio [24], pp. 919 f.

and variance of $\ln \frac{Z(T)}{Z(t)}$ as obtained by Brigo and Mercurio [24], pp. 886–889, i.e.

$$\mathbb{E}^{Q^T}\left[\ln \frac{Z(T)}{Z(t)}\Big|\mathcal{F}_t\right] = -\ln P(t,T) - \frac{1}{2}\tilde{v}(t,T)^2,$$

$$\mathbb{V}^{Q^T}\left[\ln \frac{Z(T)}{Z(t)}\Big|\mathcal{F}_t\right] = \tilde{v}(t,T)^2,$$

with $\tilde{v}(t,T), V(t,T), w(t,T)$ as in (4.16). All in all this results in

$$a(t,T) = \mathbb{E}^{Q^T}\left[\ln \frac{Z(T)}{Z(t)}\Big|\mathcal{F}_t\right] + \ln P(t,T) = -\frac{1}{2}\tilde{v}(t,T)^2,$$

$$b(t,T)^2 = \mathbb{V}^{Q^T}\left[\ln \frac{Z(T)}{Z(t)}\Big|\mathcal{F}_t\right] = \tilde{v}(t,T)^2,$$

$$c(t,T) = 0,$$

$$e(t,T) = -\frac{1}{2}\tilde{v}(t,T)^2 - \ln \tilde{K}(t,T) = \ln \frac{Z(t)}{P(t,T)\cdot K} - \frac{1}{2}\tilde{v}(t,T)^2,$$

$$f(t,T) = e(t,T) + \tilde{v}(t,T)^2,$$

and it follows that

$$U(t,T) = -Z(t)e^{c(t,T)}\Phi\left(-\frac{f(t,T)}{b(t,T)}\right) + Z(t)\tilde{K}(t,T)\Phi\left(-\frac{e(t,T)}{b(t,T)}\right)$$

$$= -Z(t)\cdot\Phi(-d_1) + P(t,T)\cdot K\cdot\Phi(-d_2)$$

with d_1 and in d_2 as in (4.16). $\qquad\qquad\qquad\qquad\qquad\qquad\qquad\square$

The volatility $\tilde{v}(t,T)$, which is used to calculate the European put option price, depends on the interest rate volatility σ_r, the state variable volatility σ_Z and the correlation ρ between the state variable process and the short rate process. As shown in the proof of Corollary 4.14 below, in the special case of constant interest rates where $\tilde{v}(t,T)$ equals $\sigma_z\sqrt{T-t}$, the formula for the European put option in Equation (4.15) reduces to the standard formula by Black and Scholes [20].

Corollary 4.14 (Special case: constant interest rates)
In the special case where $\sigma_r = 0$, $\kappa_r^P = \kappa_r^Q$ and $\theta_r^P = \theta_r^Q = r_0$, the present value in Equation (4.14) simplifies to

$$\tilde{P}(t,T) = e^{-r_0 \cdot (T-t)} - \text{BS}^{\text{Put}}(t,T,1,Z(t),\sigma_Z,r_0), \quad t \le T,$$

with

$$\text{BS}^{\text{Put}}(t,T,K,Z,\sigma_Z,r_0) := K \cdot e^{-r_0(T-t)} \cdot \Phi(-\tilde{d}_2) - Z \cdot \Phi(-\tilde{d}_1) \quad (4.17)$$

being the standard Black-Scholes formula for European put options and

$$\tilde{d}_1 := \frac{1}{\sigma_Z\sqrt{T-t}}\left(\ln\left(\frac{Z}{K}\right) + \left(r_0 + \frac{\sigma_Z^2}{2}\right) \cdot (T-t)\right),$$
$$\tilde{d}_2 := \tilde{d}_1 - \sigma_Z\sqrt{T-t}.$$

Proof. From Remark 4.8 we know that $\sigma_r = 0$, $\kappa_r^P = \kappa_r^Q$ and $\theta_r^P = \theta_r^Q = r_0$ implies $r(t) = r_0$ and $P(t,T) = e^{-r_0 \cdot (T-t)}$. With $V(t,T) = w(t,T) = 0$ and $\tilde{v}^2(t,T) = \sigma_Z^2 (T-t)$, the price of the put option in Equation (4.15) reduces to $\text{BS}^{\text{Put}}(t,T,K,Z,\sigma_Z,r_0)$ as in Equation (4.17). $\qquad \square$

In our model, the price and the spread of a risky zero coupon bond are strictly positive.

Remark 4.15 (Risky zero coupon bond price and spread)
At time $t \le T$, $t,T \in \mathcal{T}_c$, the present value and the spread of a risky zero coupon bond with maturity T are strictly positive, i.e. $\tilde{P}(t,T) > 0$ and $s(t,T) > 0$.

Proof. For $t < T$ it follows with $\Phi(x) = 1 - \Phi(-x) \; \forall x$ that

$$\begin{aligned}
\tilde{P}(t,T) &= P(t,T) - V^{\text{Put}}(t,T,1,Z(t),\sigma_Z,r(t),\theta_r^Q,\kappa_r^Q,\sigma_r,\rho) \\
&= P(t,T) - P(t,T) \cdot \Phi(-d_2) + Z \cdot \Phi(-d_1) \\
&= P(t,T) \cdot \Phi(d_2) + Z(t) \cdot \Phi(-d_1),
\end{aligned}$$

where d_1 and d_2 as in (4.16) and $\Phi(d_2), \Phi(-d_1) \in [0,1]$. It further holds that

$$P(t,T) = e^{-\ln A(t,T)} \cdot e^{-B(t,T) \cdot r(t)} > 0,$$

with $A(t,T)$, $B(t,T)$ as in Proposition 4.3. From the lognormal distribution of $Z(t)$ it follows that $Z(t) > 0$. As $\Phi(d_2)$ and $\Phi(-d_1)$ cannot be zero at the same time, we get $\tilde{P}(t,T) > 0$. For $t = T$ we have

$$\tilde{P}(T,T) = \min\left(1, Z(T)\right),$$

and the statement regarding the zero coupon bond's present value follows with $Z(T) > 0$.

Furthermore, as $V^{\text{Put}}(t,T,K,Z,\sigma_Z,r,\theta_r^Q,\kappa_r^Q,\sigma_r,\rho) = U(t,T) > 0$ for $t < T$ due to the geometric Brownian motion assumption, we have $\tilde{P}(t,T) < P(t,T)$ and therefore $s(t,T) > 0$. $\qquad\square$

4.3.3 Performance of the Bank's Risky Assets

Having determined the dynamics of the risk-free short rate and the state variables associated with the risky zero coupon bonds, we now specify the joint performance of the bank's risky assets as defined in Assumption 1. In line with our above discussions we make the following assumption:

Assumption 9 (Creditworthiness of the bank's risky assets)
For the bank's risky assets, the following holds:

(a) *The payoff of the bank's i-th risky asset of type x at its maturity $T_i^x \in \mathcal{T}_c$, $i = 1, ..., n_x$, $x \in \{CPS, OA\}$, is $\tilde{P}_i^x(T_i^x, T_i^x) := \min\left(1; Z_i^x(T_i^x)\right)$, where $Z_i^x(T_i^x)$ is the time-T_i^x-value of the asset's associated tradeable state variable. The risky asset cannot default prior to maturity and at time T_i^x it defaults if and only if $Z_i^x(T_i^x) < 1$.*

(b) *Under the real-world measure $\mathbb{P}$, the dynamics of the state variable processes Z_i^x are given by*

$$dZ_i^x(t) = \mu_i^x Z_i^x(t)dt + \sigma_i^x Z_i^x(t)dW_i^x(t), \quad \forall t \in \mathcal{T}_c, \tag{4.18}$$

for $Z_i^x(0) := Z_i^{x,0}$ and $i = 1, ..., n_x$, $x \in \{CPS, OA\}$, with $Z_i^{x,0}$, $\sigma_i^x \in \mathbb{R}^+$ being positive constants, $\mu_i^x \in \mathbb{R}$ and $(W_i^x(t))_{t \in \mathcal{T}_c}$ a standard Brownian motion under $\mathbb{P}$.

(c) *The P-dynamics of the state variable processes and the short rate are correlated according to*

$$dW_i^{CPS}(t)dW_k^{CPS}(t) = \rho_{i,k}^{CPS,CPS}dt,$$

$$dW_j^{OA}(t)dW_l^{OA}(t) = \rho_{j,l}^{OA,OA}dt,$$

$$dW_i^{CPS}(t)dW_j^{OA}(t) = \rho_{i,j}^{CPS,OA}dt,$$

$$dW_i^{CPS}(t)dW_r^{P}(t) = \rho_i^{CPS,r}dt$$

$$dW_j^{OA}(t)dW_r^{P}(t) = \rho_j^{OA,r}dt,$$

with $\rho_{i,i}^{CPS,CPS} = \rho_{j,j}^{OA,OA} = 1$ *and* $\rho_{i,k}^{CPS,CPS}, \rho_{j,l}^{OA,OA}, \rho_{i,j}^{CPS,OA} \in (-1,1)$ *for* $i \neq k$, $j \neq l$, $i,k \in \{1,...,n_{CPS}\}$, $j,l \in \{1,...,n_{OA}\}$. *Furthermore,* $\rho_i^{CPS,r}$, $\rho_j^{OA,r} \in (-1,1)$ *for* $i \in \{1,...,n_{CPS}\}$, $j = \{1,...,n_{OA}\}$ *and* $\rho^{r,r} = 1$. *The correlation parameters are chosen such that the matrix*

$$M := \begin{pmatrix}
\rho^{r,r} & \rho_1^{CPS,r} & \cdots & \rho_{n_{CPS}}^{CPS,r} & \rho_1^{OA,r} & \cdots & \rho_{n_{OA}}^{OA,r} \\
\rho_1^{CPS,r} & \rho_{1,1}^{CPS,CPS} & \cdots & \rho_{1,n_{CPS}}^{CPS,CPS} & \rho_{1,1}^{CPS,OA} & \cdots & \rho_{1,n_{OA}}^{CPS,OA} \\
\vdots & \vdots & \vdots & \vdots & \vdots & \vdots & \vdots \\
\rho_{n_{CPS}}^{CPS,r} & \rho_{n_{CPS},1}^{CPS,CPS} & \cdots & \rho_{n_{CPS},n_{CPS}}^{CPS,CPS} & \rho_{n_{CPS},1}^{CPS,OA} & \cdots & \rho_{n_{CPS},n_{OA}}^{CPS,OA} \\
\rho_1^{OA,r} & \rho_{1,1}^{CPS,OA} & \cdots & \rho_{n_{CPS},1}^{CPS,OA} & \rho_{1,1}^{OA,OA} & \cdots & \rho_{1,n_{OA}}^{OA,OA} \\
\vdots & \vdots & \vdots & \vdots & \vdots & \vdots & \vdots \\
\rho_{n_{OA}}^{OA,r} & \rho_{1,n_{OA}}^{CPS,OA} & \cdots & \rho_{n_{CPS},n_{OA}}^{CPS,OA} & \rho_{n_{OA},1}^{OA,OA} & \cdots & \rho_{n_{OA},n_{OA}}^{OA,OA}
\end{pmatrix}$$

is positive definite.

With the help of the Cholesky decomposition, the joint dynamics of the short rate and the state variable processes can be expressed in terms of independent Brownian motions.

Remark 4.16 (Joint P-dynamics of the bank's risk drivers)
Under the real-world measure $\mathbb{P}$, *the joint dynamics of the short rate* r *and the state variables* Z_i^x *of the bank's risky assets,* $i = 1,...,n_x$, $x \in \{CPS, OA\}$, *can be written as*

$$d\tilde{Y}(t) = \mu_{\tilde{Y}}^{P}(t)dt + \Sigma_{\tilde{Y}}(t)dW_{\tilde{Y}}^{P}(t), \quad \forall t \in \mathcal{T}_c, \tag{4.19}$$

where

$$\tilde{Y}(t) := \left(r(t), Z_1^{CPS}(t), ..., Z_{n_{CPS}}^{CPS}(t), Z_1^{OA}(t), ..., Z_{n_{OA}}^{OA}(t) \right)^\top \in \mathbb{R}^{\tilde{n}},$$

$$\mu_{\tilde{Y}}^P(t) := \left(\mu_r^P(t), \mu_1^{CPS}(t), ..., \mu_{n_{CPS}}^{CPS}(t), \mu_1^{OA}(t), ..., \mu_{n_{OA}}^{OA}(t) \right)^\top \in \mathbb{R}^{\tilde{n}},$$

with $\tilde{n} := n_{CPS} + n_{OA} + 1$, $r(0) := r_0$, $Z_i^x(0) := Z_i^{x,0}$ and

$$\mu_r^P(t) := \kappa_r^P \left(\theta_r^P - r(t) \right), \quad \mu_i^x(t) := \mu_i^x Z_i^x(t).$$

Furthermore, $\Sigma_{\tilde{Y}}(t) \in \mathbb{R}^{\tilde{n} \times \tilde{n}}$ is the lower triangular matrix from the Cholesky decomposition of the variance-covariance matrix $\Sigma_{\tilde{Y}}^{VC}(t) := \Sigma_M(t) \cdot M \cdot \Sigma_M(t)$ with the diagonal matrix $\Sigma_M(t)$ being defined by

$$\Sigma_M(t) := \mathrm{diag}(\sigma_r, \tilde{\sigma}_1^{CPS}(t), ..., \tilde{\sigma}_{n_{CPS}}^{CPS}(t), \tilde{\sigma}_1^{OA}(t), ..., \tilde{\sigma}_{n_{OA}}^{OA}(t)) \in \mathbb{R}^{\tilde{n} \times \tilde{n}}$$

and $\tilde{\sigma}_i^x(t) := \sigma_i^x Z_i^x(t)$, i.e. $\Sigma_{\tilde{Y}}(t) \cdot \Sigma_{\tilde{Y}}(t)^\top = \Sigma_{\tilde{Y}}^{VC}(t)$. The parameters r_0, κ_r^P, θ_r^P, σ_r are as in Assumption 6, $Z_i^{x,0}$, σ_i^x, μ_i^x, M as in Assumption 9 and $(W_{\tilde{Y}}^P(t))_{t \in \mathcal{T}_c}$ is a $\tilde{n}$-dimensional standard Brownian motion under $\mathbb{P}$.

With respect to the joint dynamics of the bank's risk drivers under $\mathbb{Q}$, we make the following assumption:

Assumption 10 (Joint $\mathbb{Q}$-dynamics of the bank's risk drivers)
The Radon-Nikodym derivative of the risk-neutral measure $\mathbb{Q}$ with respect to $\mathbb{P}$, restricted to $\mathcal{F}_t$, is given by

$$\left. \frac{d\mathbb{Q}}{d\mathbb{P}} \right|_{\mathcal{F}_t} = \exp \left(-\frac{1}{2} \int_0^t \|\gamma_{\tilde{Y}}(s)\|^2 ds - \int_0^t \gamma_{\tilde{Y}}^\top(s) dW_{\tilde{Y}}^P(s) \right),$$

where

$$\gamma_{\tilde{Y}}(s) := \left(\Sigma_{\tilde{Y}}(t) \right)^{-1} \cdot \left(\mu_{\tilde{Y}}^P(t) - \mu_{\tilde{Y}}^Q(t) \right) \in \mathbb{R}^{\tilde{n}}, \tag{4.20}$$

and

$$\mu_{\tilde{Y}}^Q(t) := \left(\mu_r^Q(t), \tilde{\mu}_1^{CPS}(t), ..., \tilde{\mu}_{n_{CPS}}^{CPS}(t), \tilde{\mu}_1^{OA}(t), ..., \tilde{\mu}_{n_{OA}}^{OA}(t) \right)^\top \in \mathbb{R}^{\tilde{n}}.$$

Furthermore, $\mu_r^Q(t) := \kappa_r^Q \left(\theta_r^Q - r(t) \right)$ *and* $\tilde{\mu}_i^x(t) := r(t) Z_i^x(t)$ *for* $i = 1, ..., n_x$ *and* $x \in \{CPS, OA\}$.

Given Assumptions 9 and 10, the resulting $\mathbb{Q}$-dynamics can be easily derived.

Proposition 4.17 (Joint $\mathbb{Q}$-dynamics of the bank's risk drivers)
Under the risk-neutral measure $\mathbb{Q}$, the joint dynamics of the short rate r and the state variables Z_i^x, $i = 1, ..., n_x$, $x \in \{CPS, OA\}$, are given by

$$d\tilde{Y}(t) = \mu_{\tilde{Y}}^Q(t)dt + \Sigma_{\tilde{Y}}(t)dW_{\tilde{Y}}^Q(t), \quad \forall t \in \mathcal{T}_c,$$

where $\tilde{Y}(t)$, $\sigma_{\tilde{Y}}(t)$ are as in Remark 4.16, $\mu_{\tilde{Y}}^Q(t)$ is as in Assumption 10 and $(W_{\tilde{Y}}^Q(t))_{t \in \mathcal{T}_c}$ is a $\tilde{n}$-dimensional standard Brownian motion under $\mathbb{Q}$.

Proof. From Assumptions 9, 10 and Remark 4.16 it follows with Girsanov's theorem, cf. Zagst [168], pp. 33–35, that the stochastic process

$$dW_{\tilde{Y}}^Q(t) = \gamma_{\tilde{Y}}(t)dt + dW_{\tilde{Y}}^P(t)$$

is a $\tilde{n}$-dimensional standard Brownian motion under $\mathbb{Q}$. Substituting $dW_{\tilde{Y}}^P(t)$ in Equation (4.19) and setting $\gamma_{\tilde{Y}}(s)$ as in Equation (4.20) we get

$$
\begin{aligned}
d\tilde{Y}(t) &= \mu_{\tilde{Y}}^P(t)dt - \Sigma_{\tilde{Y}}(t)\gamma_{\tilde{Y}}(t)dt + \Sigma_{\tilde{Y}}(t)dW_{\tilde{Y}}^Q(t) \\
&= \mu_{\tilde{Y}}^P(t)dt - \Sigma_{\tilde{Y}}(t)\left(\Sigma_{\tilde{Y}}(t)\right)^{-1} \cdot \left(\mu_{\tilde{Y}}^P(t) - \mu_{\tilde{Y}}^Q(t)\right)dt + \Sigma_{\tilde{Y}}(t)dW_{\tilde{Y}}^Q(t) \\
&= \mu_{\tilde{Y}}^Q(t)dt + \Sigma_{\tilde{Y}}(t)dW_{\tilde{Y}}^Q.
\end{aligned}
$$

$\square$

We conclude this section by introducing some additional notation which will be needed later on.

Realized cash flow per unit of risky asset. At time $t \in \mathcal{T}_c$, the realized cash flow from one unit of the risky asset $i \in \{1, ..., n_x\}$ of type $x \in \{CPS, OA\}$ is denoted by

$$\tilde{F}_{i,U}^x(t) := \begin{cases} \min\left(1; Z_i^x(T_i^x)\right), & \text{if } t = T_i^x, \\ 0, & \text{otherwise.} \end{cases}$$

Realized present value per unit of risky asset. At time $t \in \mathcal{T}_c$, the present value of one unit of the risky asset $i \in \{1, ..., n_x\}$ of type $x \in \{CPS, OA\}$, is denoted by

$$V_{i,U}^x(t) := \begin{cases} \tilde{P}(t, T_i^x), & \text{if } t \leq T_i^x, \\ 0, & \text{otherwise,} \end{cases}$$

Due to the zero coupon bond assumption, there is no distinction between cash flows from nominal repayments and interest rate payments.

4.4 Funding and Liquidity

Bank survival strongly depends on the bank's ability to obtain funding when needed. In the following, we specify the bank's and the cover pool's liquidity situation and the conditions under which funding is available in our model (Sections 4.4.2 and 4.4.3). As the maintenance of the legal cover requirements plays an important role in this context, we start with the definition of the bank's matching cover calculations (Section 4.4.1).

4.4.1 Matching Cover Calculations

One of the key features of the Pfandbrief is its mandatory overcollateralization. There are four cover requirements which have to be maintained by the bank at all times, cf. Section 2.1: the nominal cover, the excess cover, the excess cover under stress and the 180-day liquidity buffer. To ensure that that the necessary cover is given, the Pfandbrief bank performs regular matching cover calculations. As cover requirements are not actively maintained any more by the cover pool administrator, matching cover calculations only take place until the time of bank default.

Assumption 11 (The bank's matching cover calculations)
Matching cover calculations are performed according to the following rules:

(a) The nominal cover, the excess cover, the excess cover under stress and the 180-day liquidity buffer are in scope.

(b) *Cover requirements refer to strategic cover pool assets. Cover pool cash inflows and Pfandbrief payments due at the time of calculation are not taken into consideration.*

(c) *The risk-free interest rate curve is used for discounting when calculating the excess cover.*

(d) *The excess cover under stress is determined by shifting the risk-free curve by 250 bp up and down and setting resulting negative interest rates to zero.*

The PfandBG requires the excess cover under stress to be reviewed once a week, while all other cover requirements have to be checked on a daily basis. Assuming that all four cover requirements are in scope of each matching cover calculation is therefore a reasonable approximation of the legal provisions, especially for simulation time steps larger than one week. The second assumption, which states that cover requirements refer to strategic cover pool assets, is motivated by the fact that in our model strategic cover pool assets represent the bank's Pfandbrief business while liquid cover pool assets are only used for liquidity steering.[10] As there is no dedicated one-to-one relationship between cover pool cash inflows and Pfandbrief payments, cash flows due at the time of matching cover calculations are not considered in the calculations. The second assumption also implies that all strategic cover pool assets are eligible as cover. In the case of Public Pfandbriefe where rating restrictions apply in some cases, cf. Section 2.1, this is an approximation to avoid additional asset liability management related assumptions in the case of rating deteriorations which result in single assets becoming ineligible for the cover pool. For Mortgage Pfandbriefe, on which we concentrate in the following, this implication is of lower relevance as there are no explicit rating restrictions. Assuming that the risk-free curve is used for discounting is a viable approximation of the legal provisions which require the corresponding present values to be calculated by discounting expected future cash flows with the currency-specific swap curve. Currency stresses are not relevant in our setup as all assets are denoted in euros. The assumption regarding the shocks to be applied when calculating the excess cover under stress is in line with the legal provisions according to which the bank can choose between a static approach, a dynamic approach and a method based on the bank's internal risk model.[11] Under the static approach, which is the least

---

[10]For more details, see also Section 4.6 below.
[11]More details can be found in the PfandBarwertV, see vdp [157].

complex one, present values are calculated by shifting the reference curve by
250 basis points up and down and setting resulting negative interest rates to
zero.

Having specified the principles according to which matching cover calculations
are performed, we now translate the legal cover requirements into formulas.
The nominal cover requirement states that the nominal of the cover pool,
i.e. the strategic cover pool assets in our case, must be equal to or exceed
the nominal of outstanding Pfandbriefe.

Observation 4.18 (Nominal cover)
At time $t \in \mathcal{T}$, the nominal cover requirement is given by

$$\tilde{N}^{CPS}(t) \geq \tilde{N}^{PB}(t), \tag{C1}$$

with $\tilde{N}^x(t) := \sum_{i \in \mathcal{N}_x, T_i^x > t} N_i^x(t)$ and $\mathcal{N}_x := \{1, ..., n_x\}$ for $x \in \{CPS, PB\}$.

Note that, in line with Assumption 11, the nominals $\tilde{N}^x(t)$ do not account
for time-t cash inflows and payments.

The provisions regarding the excess cover and the excess cover under stress
require the net present value of cover pool assets to exceed the net present
value of outstanding Pfandbriefe by at least 2% and this must still hold
under the above defined interest rate stresses.

Observation 4.19 (Excess cover and excess cover under stress)
*At time $t \in \mathcal{T}$, the excess cover and the excess cover under stress requirements
are given by*

$$\tilde{V}_{*y}^{CPS}(t) \geq 1.02 \cdot \tilde{V}_{*y}^{PB}(t), \quad y \in \{b, u, d\}, \tag{C2}$$

*with $\tilde{V}_{*y}^x(t) := \sum_{s \in \mathcal{T}_x, s > t} F_t^x(s) \cdot d_{*y}(t, s)$ and $F_t^x(s) := \sum_{i \in \mathcal{N}_x, T_i^x = s} N_i^x(t)$,
where $d_{*y}(t, s) := e^{-\max(0; r(t,s) + \Delta_y) \cdot (s-t)}$ and $y \in \{b, u, d\}$, $x \in \{CPS, PB\}$.
Furthermore, $\mathcal{T}_x := \{T_1^x, ..., T_{n_x}^x\}$, $\Delta_b := 0$, $\Delta_u := 0.025$ and $\Delta_d := -0.025$.*

Here, $y = b$ corresponds to the excess cover requirement, while $y \in \{u, d\}$
represents the corresponding up and down shocks of the excess cover under
stress requirement. By construction, the net present values $\tilde{V}_{*y}^x(t)$ do not
consider time-t cash inflows and payments. Negative interest rates resulting
from the shocks are floored at zero before used for discounting. Note that

$F_t^x(s)$ and $\tilde{V}_{*y}^x(t)$ account for our special zero coupon bonds setting, where cash flows solely arise at maturity.

Finally, the 180-day liquidity buffer states that the maximum cumulative net cash outflow occurring within the next 180 days must be covered by highly liquid assets and that the cumulative difference between cash inflows from the cover pool and scheduled Pfandbrief payments needs to be determined with daily granularity.

Observation 4.20 (180-day liquidity buffer)
At time $t \in \mathcal{T}$, the 180-day liquidity buffer requirement is given by

$$F_{cum}^{CPS}(t,s) \geq F_{cum}^{PB}(t,s), \quad \forall s \in \mathcal{T}_{CPS}^{PB}, \quad t < s \leq t + 0.5, \qquad (C3)$$

with $F_{cum}^x(t,s) := \sum_{i \in \mathcal{N}_x, t < T_i^x \leq s} N_i^x(t)$ for $x \in \{CPS, PB\}$. Furthermore, $\mathcal{T}_{CPS}^{PB} := \mathcal{T}_{CPS} \cup \mathcal{T}_{PB}$.

In the following, we require that all cover requirements are fulfilled at time t_0.

Assumption 12 (Cover requirements in t_0)
The legal cover requirements (C1)–(C3) are fulfilled in t_0.

As soon as one of the legal cover requirements (C1)–(C3) is breached, the bank has to take action. Its strategies in this context are specified in Section 4.6 below.

4.4.2 Bank Funding

Under the run-off assumption, which implies that no new business is made, bank funding is mainly needed to overcome temporary cash shortages resulting from asset liability mismatches and the need to maintain cover requirements.

Definition 4.21 (Bank funding need)
The bank's funding need at time $t \in \mathcal{T}$ is given by

$$G_B^C(t) := \begin{cases} \max(0; F_B^L(t) + CR(t) - \tilde{F}_B^A(t)), & \text{if } t < T_{\max}, \ t \le \tau, \\ 0, & \text{otherwise.} \end{cases}$$

with $CR(t)$ being the amount of cash needed to reestablish the legal cover requirements as specified in Proposition 4.41 below and

$$F_B^L(t) := F^{PB}(t) + F^{OL}(t) + N_B^{LL}(t),$$
$$\tilde{F}_B^A(t) := \tilde{F}^{CPS}(t) + \tilde{F}^{OA}(t) + N^{CPL}(t),$$
$$F^x(t) := \sum_{i \in \mathcal{N}_x, T_i^x = t} N_i^x(t_0),$$
$$\tilde{F}^y(t) := \sum_{i=1}^{n_y} N_i^y(t) \cdot \tilde{F}_{i,U}^y(t),$$

for $x \in \{PB, OL\}$ and $y \in \{CPS, OA\}$.

According to the above definition, a bank funding need arises when the time-t asset cash inflows are not sufficient to fulfil payment obligations and to maintain cover requirements. As the simulation stops at time $T_{\max}$ and the balance sheet is liquidated, funding aspects are not relevant any more at that time and $G_B^C(T_{\max}) = 0$.

As already discussed in Section 3.3.1, access to funding highly depends on the market environment and the bank's specific situation. The availability of funding through a collateralized liquidity line was found to be a reasonable assumption as it combines the characteristics of secured funding as a reliable funding source with the ones of a liquidity line where cash can be raised at predefined conditions, avoiding the need for an explicit modelling of funding costs. To account for the fact that liquidity lines may be withdrawn when conditions deteriorate, we stipulate that the provider of the liquidity line has the option to decide whether or not to prolong the granted funding at each point in time. Alternatively, one could assume that further Pfandbriefe (another form of secured funding) are issued to overcome temporary funding needs. This would, however, raise non-trivial questions regarding the fair coupon and the maturity of such instruments and might not be a realistic option in case of a funding need caused by a breach of cover requirements.

Assumption 13 (The bank's liquidity line)
At time $t \in \mathcal{T}$, $t \leq \tau$, the bank can obtain funding from a collateralized liquidity line according to the following conditions:

(a) The bank has to pledge assets as collateral. It can use other assets for this purpose and, if cover requirements are still fulfilled for the remaining part of the cover pool, strategic cover pool assets. The maturity of pledged assets must be later than time t.

(b) Assets are pledged at a haircut to their present value. For other assets, this haircut is $h_{PL}^{OA} \in [0,1]$ and for strategic cover pool assets it is $h_{PL}^{CPS} \in [0,1]$.

(c) The funding has to be paid back in $t + \Delta \in \mathcal{T}$ and the amount to be paid is given by

$$N_B^{LL}(t+\Delta) := x \cdot P(t, t+\Delta)^{-1},$$

with $x > 0$ being the amount borrowed at time t.

(d) Pledged assets remain on the bank's balance sheet but are marked as pledged until the funding has been paid back in $t + \Delta$. Pledged strategic cover pool assets must be temporarily removed (deregistered) from the cover pool but can be registered again once the funding has been paid back.

(e) In case of a bank default, the provider of the liquidity line has a priority claim on the pledged assets, including the cash inflows arising from these assets at the time of default. In addition, he also has an unsecured claim against the bank's general insolvency estate.

(f) No funding is granted in $T_{\max}$.

The liquidity line as specified by Assumption 13 is similar to the funding option in Liang et al [114], who assume that the bank can under certain conditions overcome liquidity needs by entering into collateralized funding activities with risky assets being pledged at a haircut to their present value. This haircut is in line with secured funding practice and protects the counterpart of the funding transaction from potential losses due to uncertainties regarding the future realisable sale price of pledged assets caused by price volatility and potential liquidation discounts.[12] As opposed to Liang et al who define one fixed repayment date in the (potentially far) future, namely the maturity of the bank's long term debt, the funding in our model is only granted for a short time period of length Δ. By assuming that funding needs to be renewed in $t + \Delta$, we account for potential deteriorations of collateral quality which may result in the necessity to post additional collateral over time, and we also take into consideration the possibility of liquidity line withdrawals.[13] In addition, a periodic renewal of funding as in our model leaves the bank with the flexibility to repay the borrowed amounts once they are not needed any more. Following Liang et al [114] we further assume that the provider of the collateralized funding has a priority claim on the pledged assets. In our setup, this implies that, as long as the funding has not been paid back, pledged strategic cover pool assets have to be removed from the cover pool. This can be interpreted as a temporary reduction of voluntary overcollateralization to ease funding pressure. In addition to the priority claim on pledged assets, our liquidity provider also has an unsecured claim against the general insolvency estate. In Liang et al's paper this claim is of zero value due to their zero recovery assumption. As the simulation stops in $T_{\max}$, no more funding is available at that time.

Figure 4.3 illustrates how funding activities change the composition of the bank's balance sheet. For risky assets, funding activities result in a distinction of pledged and unpledged assets. On the liabilitiy side, an additional position appears: the bank's liquidity line.

In the context of asset pledging we introduce some more notation.

Time-t nominal of pledged and non-pledged assets The nominals of pledged assets at time t, $t \in \mathcal{T}$, which stem from collateralized funding activities at time $t - \Delta$, are denoted by

$$N_{i,PL}^{CPS}(t), \quad N_{j,PL}^{OA}(t),$$

---

[12]The realisable sale price becomes important once the bank is not able any more to repay the borrowed amount, leaving the funding provider with the liquidation proceeds of the pledged assets.

[13]If the bank's economic situation (represented by the present value of assets available for pledging in this case) deteriorates, the provider of the liquidity line might not be willing any more to prolong the funding.

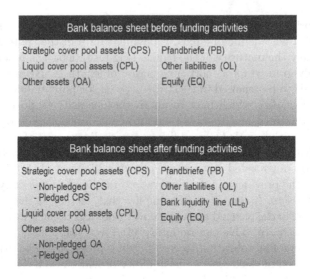

Figure 4.3: Balance sheet before and after bank funding activities.

for $i = 1, ..., n_{CPS}$ and $j = 1, ..., n_{OA}$, and the corresponding nominals of non-pledged assets are given by

$$N_{i,NPL}^{CPS}(t), \quad N_{j,NPL}^{OA}(t),$$

with $N_{k,PL}^{x}(t) + N_{k,NPL}^{x}(t) = N_{k}^{x}(t)$, $x \in \{CPS, OA\}$ and $k = 1, ..., n_x$. Furthermore, we set

$$N_y^x(t) := \sum_{i=1}^{n_x} N_{i,y}^x(t), \quad y \in \{PL, NPL\}.$$

For $t \in \mathcal{T}, t > t_0$, all these nominals are $\mathcal{F}_{t-\Delta}$-measurable.

In the following, we assume that the bank's liquidity line has not been drawn prior to time t_0.

Assumption 14 (The bank's liquidity line in t_0)
At time t_0, it holds that

$$N_B^{LL}(t_0) = N_{i,PL}^{CPS}(t_0) = N_{j,PL}^{OA}(t_0) = N_{PL}^{CPS}(t_0) = N_{PL}^{OA}(t_0) = 0.$$

Furthermore, $N_{i,NPL}^x(t_0) = N_i^x(t_0)$ and $N_{NPL}^x(t_0) = N^x(t_0)$ for $i = 1, ..., n_x$ and $x \in \{CPS, OA\}$.

The higher the asset present values and the lower the haircuts, the more funding can be raised. In the special case of strategic cover pool assets, the maximum amount of funding obtainable also depends on the extent to which cover requirements are fulfilled. If one of the cover requirements is already breached when funding needs to be raised, no funding from pledging strategic cover pool assets is available at all.

Observation 4.22 (Maximum bank funding)
Under the assumption that asset pledging within each of the two asset classes is done pro rata, the maximum amount of funding available to the bank at time $t \in \mathcal{T}$ is given by

$$L_B^{\max}(t) := L_{OA}^{\max}(t) + L_{CPS}^{\max}(t) \cdot z_B^{CPS}(t),$$

with

$$L_x^{\max}(t) := \begin{cases} (1 - h_{PL}^x) \cdot \tilde{V}^x(t), & \text{if } t < T_{\max}, \ t \leq \tau, \\ 0, & \text{otherwise,} \end{cases}$$

and

$$\tilde{V}^x(t) := \sum_{i \in \mathcal{N}_x, T_i^x > t} N_i^x(t) \cdot V_{i,U}^x(t),$$

for $x \in \{CPS, OA\}$. Furthermore, the fraction $z_B^{CPS}(t)$ is defined by

$$z_B^{CPS}(t) := \sup \{z \in [0,1] \mid (4.21) - (4.23) \text{ hold}\}$$

where $\sup \emptyset := 0$ and

$$\tilde{N}^{CPS}(t) \cdot (1 - z) \geq \tilde{N}^{PB}(t), \tag{4.21}$$

$$\tilde{V}_{*y}^{CPS}(t) \cdot (1 - z) \geq 1.02 \cdot \tilde{V}_{*y}^{PB}(t), \quad y \in \{b, u, d\}, \tag{4.22}$$

$$F_{cum}^{CPS}(t, s) \cdot (1 - z) \geq F_{cum}^{PB}(t, s), \quad \forall s \in \mathcal{T}_{CPS}^{PB}, \ t < s \leq t + 0.5. \tag{4.23}$$

By definition, $L_{OA}^{\max}(t)$ corresponds to the maximum amount of funding which can be raised by pledging other assets at a haircut h_{PL}^{OA} to their present value. Under the assumption that strategic cover pool assets can be pledged without restrictions, $L_{CPS}^{\max}(t)$ is the maximum amount of funding obtainable by pledging these assets at a haircut h_{PL}^{CPS}. As the cover requirements (C1)–(C3) also need to be fulfilled after strategic cover pool assets have been pledged and (4.21)–(4.23) represent the corresponding cover requirements post asset pledging, $z_B^{CPS}(t)$ is per definition the maximum fraction of strategic cover pool assets which can be removed from the cover pool without breaching one of the cover requirements. In line with Assumptions 11 and 13, cash flows realized at time t are not considered in the calculations.

4.4.3 Cover Pool Funding

Once the bank has defaulted, the cover pool administrator becomes responsible for making contractual payments to Pfandbrief holders. Under this scenario, a cover pool funding need arises when cover pool cash inflows are not sufficient to make contractual payments.

Definition 4.23 (Cover pool funding need)
The cover pool's funding need at time $t \in \mathcal{T}$ is given by

$$G_P^C(t) := \begin{cases} \max(0; F_P^L(t) - \tilde{F}_P^A(t)), & \text{if } t < T_{\max}, \ \tau \leq t \leq \tau^*, \\ 0, & \text{otherwise,} \end{cases}$$

with

$$F_P^L(t) := F^{PB}(t) + N_P^{LL}(t),$$
$$\tilde{F}_P^A(t) := \tilde{F}_{NPL}^{CPS}(t) + N^{CPL}(t),$$
$$F^{PB}(t) := \sum_{k \in \mathcal{N}_{PB}, T_k^{PB} = t} N_k^{PB}(t),$$
$$\tilde{F}_{NPL}^{CPS}(t) := \sum_{i=1}^{n_{CPS}} N_{i,NPL}^{CPS}(t) \cdot \tilde{F}_{i,U}^{CPS}(t).$$

According to the above definition, only non-pledged strategic cover pool assets are considered when determining the cover pool's cash inflows $\tilde{F}_P^A(t)$. This accounts for the fact that, at the time τ of bank default, pledged strategic cover pool assets cannot be separated at the benefit of the Pfandbrief holder, cf. Assumption 13. At later times $t > \tau$, there are no more pledged assets, i.e. $\tilde{F}_{NPL}^{CPS}(t) = \tilde{F}^{CPS}(t)$. As in the bank's case, we set $G_P^C(T_{\max}) = 0$.

The cover pool administrator has several options to raise funding when needed, cf. Section 2.1. He can enter into refinancing activities or sell cover pool assets for this purpose. In the following, we assume that there is a liquidity line as in the bank's case, but the conditions under which the funding is granted are different.

Assumption 15 (The cover pool's liquidity line)
At time $t \in \mathcal{T}$, $\tau \leq t \leq \tau^*$, the cover pool administrator can obtain funding from a liquidity line under the following conditions:

- *Funding is only granted if the nominal cover, the excess cover and the excess cover under stress are still fulfilled after the additional liability from the liquidity line is added to the balance sheet.*

- *The funding has to be paid back in $t + \Delta \in \mathcal{T}$ and the amount to be repaid is given by*

$$N_P^{LL}(t + \Delta) := x \cdot P(t, t + \Delta)^{-1},$$

with $x > 0$ being the amount borrowed at time t.

- *In case of a of cover pool default, the liquidity line provider's claim on the cover pool ranks pari passu with the one of Pfandbrief holders.*

- *No funding is granted in $T_{\max}$.*

As opposed to the bank's case, there is no asset pledging now. The only assets available to the cover pool administrator are cover pool assets and using these assets as collateral would explicitly shift the priority claim from the Pfandbrief holders to the provider of the liquidity line, which is against the legal provisions. Instead, we assume that funding can be obtained as long as the legal cover requirements (C1) and (C2) are fulfilled after the claim of the liquidity line has been added to the liability side. The assumption that the cover pool's liquidity line ranks pari passu with the creditors of outstanding Pfandbriefe is motivated by the argument that it would be rather unlikely to find someone willing to lend money to a cover pool under funding pressure if the associated claim was subordinate.

Figure 4.4 illustrates how cover pool funding activities change the composition of the balance sheet. While the asset side remains unchanged, a new position appears on the liability side: the cover pool's liquidity line.

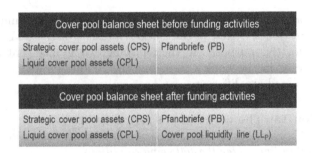

Figure 4.4: Balance sheet before and after cover pool funding activities.

In the following, we assume that no funding has been raised by the cover pool administrator prior to the time of bank default.

Assumption 16 (Cover pool liquidity line prior to bank default)
At time $t \in \mathcal{T}$, $t \leq \tau$, it holds that $N_P^{LL}(t) = 0$.

The maximum amount of funding available to the cover pool administrator depends on the extent to which the cover requirements (C1) and (C2) are fulfilled. If one of these requirements is already breached when the funding need arises, no funding can be obtained.

Observation 4.24 (Maximum cover pool funding)
The maximum amount of funding available to the cover pool administrator at time $t \in \mathcal{T}$ is given by

$$L_P^{\max}(t) := \begin{cases} \sup\left\{ z \in \mathbb{R}_0^+ \mid (4.24)\&(4.25) \text{ hold} \right\}, & \text{if } t < T_{\max}, \ \tau \leq t \leq \tau^*, \\ 0, & \text{otherwise}, \end{cases}$$

where $\sup \emptyset := 0$ *and*

$$\tilde{N}_{NPL}^{CPS}(t) \geq \tilde{N}^{PB}(t) + N_z^{LL}(t), \tag{4.24}$$

$$\tilde{V}_{*y,NPL}^{CPS}(t) \geq 1.02 \cdot \tilde{V}_{*y}^{x}(t), \quad y \in \{b, u, d\}, \tag{4.25}$$

with

$$N_z^{LL}(t) := z \cdot P(t, t+\Delta)^{-1},$$

$$\tilde{N}_{NPL}^{CPS}(t) := \sum_{i \in \mathcal{N}_{CPS}, T_i^{CPS} > t} N_{i,NPL}^{CPS}(t),$$

$$\tilde{V}_{*y,NPL}^{CPS}(t) := \sum_{s \in \mathcal{T}_{CPS}, s > t} F_{t,NPL}^{CPS}(s) \cdot d_{*y}(t, s),$$

$$F_{t,NPL}^{CPS}(s) := \sum_{i \in \mathcal{N}_{CPS}, T_i^{CPS} = s} N_{i,NPL}^{CPS}(t),$$

$$\tilde{V}_{*y}^{x}(t) := \tilde{V}_{*y}^{PB}(t) + N_z^{LL}(t) \cdot d_{*y}(t, t+\Delta).$$

As the cover requirements (C1) and (C2) still need to be fulfilled post funding and (4.24) and (4.25) represent the corresponding cover requirements when the additional claim from the liquidity line is added, $L_P^{\max}(t)$ is per definition the maximum amount of funding which can be raised. In line with Assumption 11, cash flows realized at time t are not accounted for. As compared to the bank's case, we now only consider non-pledged strategic cover pool assets when calculating the cover requirements post funding.

4.5 Specification of Default Events and Liquidation Payments

In this section we specify the events of bank and cover pool default (Sections 4.5.1 and 4.5.2) and derive the associated liquidation payments (Section 4.5.3). We do so by distinguishing two different default triggering events: overindebtedness and illiquidity. While overindebtedness is caused by a deterioration of asset quality, illiquidity stems from the inability to raise enough funding to fulfil payment obligations. Both default events are influenced by the market environment and our liquidation payments take into consideration the Pfandbrief-specific priority of payments. The default modelling requirements from Table 2.2 are therefore fulfilled.

4.5.1 Bank Default

As outlined in Section 3.2, the majority of structural credit risk models assume that default is triggered once the bank's asset value falls below a given barrier, which can be interpreted as some kind of overindebtedness. Based on our previous discussions we define bank overindebtedness as follows.

Definition 4.25 (Bank overindebtedness)
The bank is said to be overindebted at time $t \in \mathcal{T}$, $t \leq \tau$, if

$$V_B^O(t) < B_B^O(t), \tag{OB}$$

with

$$V_B^O(t) := V^{CPS}(t) + V^{OA}(t) + N^{CPL}(t),$$

$$B_B^O(t) := B_{PB}^O(t) + B_{OL}^O(t) + N_B^{LL}(t),$$

$$V^x(t) := \sum_{i=1}^{n_x} N_i^x(t) \cdot V_{i,U}^x(t), \quad x \in \{CPS, OA\},$$

$$B_y^O(t) := \sum_{i=1}^{n_y} N_i^y(t) \cdot f_O(t, T_i^y) \cdot d(t, T_i^y), \quad y \in \{PB, OL\},$$

and

$$f_O(t, s) := \begin{cases} M_{ST}, & \text{if } t \leq s \leq t + T_{ST}, \\ M_{LT}, & \text{if } s \geq t + T_{LT}, \\ M_{ST} + (s - t - T_{ST}) \cdot \frac{M_{LT} - M_{ST}}{T_{LT} - T_{ST}}, & \text{if } t + T_{ST} < s < t + T_{LT}, \\ 0, & \text{otherwise}, \end{cases}$$

$$d(t, T_i^y) := \begin{cases} P(t, T_i^y), & \text{if } t \leq T_i^y, \\ 0, & \text{otherwise}, \end{cases}$$

where $0 \leq M_{LT} \leq M_{ST} \leq 1$ and $0 \leq T_{ST} < T_{LT}$.

The barrier $B_B^O(t)$ accounts for the bank's outstanding debt at time t and includes debt repayments due at that time. As in the barrier in Equation (3.2), future debt repayments are discounted back to time t. In addition, the contribution of future payments further depends on their actual timing, which is inspired by the default barrier used in Moody's KMV, see Equation (3.3). While we account for short term payments in the next T_{ST} years with a factor M_{ST}, long-term payments due in T_{LT} years or later are multiplied by a factor $M_{LT} \leq M_{ST}$. To avoid artificial jumps in the barrier, we use a linear interpolation for mid-term payments between T_{ST} and T_{LT} years.

The choice of the barrier $B_B^O(t)$ is a very central assumption. Its specification is not straightforward and can have a large impact on model results. As we will see in Section 6.3 below, simulation results obtained from an exemplary model calibration indicate that the risk of an underestimation of bank default probabilities due to a wrong barrier specification is mitigated when illiquidity is also considered as default reason.

Assumption 17 (Bank overindebtedness in t_0)
The bank is not overindebted at time t_0.

Following Liang et al [114], we also consider illiquidity to be a potential default reason, in addition to overindebtedness.

Definition 4.26 (Bank illiquidity)
The bank is said to be illiquid at time $t \in \mathcal{T}$, $t \leq \tau$, if

$$V_B^O(t) \geq B_B^O(t) \quad \text{and} \quad G_B^C(t) > L_B^{\max}(t). \tag{IB}$$

According to the above definition, the bank is illiquid if it is not overindebted but its funding need $G_B^C(t)$ is bigger than the maximum amount of funding $L_B^{max}(t)$ obtainable. As $G_B^C(T_{\max}) = 0$ by definition, the bank cannot be illiquid in $T_{\max}$. Due to Assumptions 1, 12 and 14, it is also not illiquid in t_0 as $G_B^C(t_0) = \max(0; F_B^L(t_0) + CR(t_0) - \tilde{F}_B^A(t_0)) = \max(0; -N^{CPL}(t_0)) = 0$.

In the following, we assume that bank default occurs once the exogenously given solvency barrier is hit or the current liquidity need cannot be overcome by raising additional funding.

Assumption 18 (Bank default)
The time τ of bank default is given by

$$\tau := \inf \{t \in \mathcal{T} \mid (OB) \text{ or } (IB) \text{ holds}\},$$

with $\inf \emptyset := \infty$ *and* $\tau = \infty$ *being interpreted as no bank default occurring until time* $T_{\max}$.

As (OB) and (IB) cannot hold at the same time, there is always a well-defined reason for bank default, see Figure 4.5 below. Note that our distinction of bank default scenarios corresponds to the one in Liang et al's special case where t is a rollover date and equals the first time of a bank run (cases 4, 5 and 6 in Figure 3.7).[14]

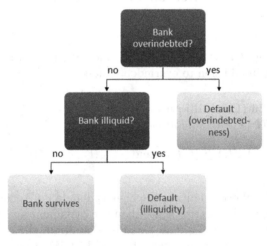

Figure 4.5: Bank default reasons at time $t \leq \tau$.

Regarding the consequences of bank default, we make the following assumption:

---

[14]In the description of the model by Liang et al, 'solvent' corresponds to what we refer to as 'not overindebted'.

Assumption 19 (Consequences of bank default)
The consequences of a bank default at time $\tau \in \mathcal{T}$ are as follows:

(a) The fate of the cover pool and outstanding Pfandbriefe depends on whether or not the cover pool survives the event of bank default. In case it survives, a standalone cover pool under the administration of a cover pool administrator is created and outstanding Pfandbriefe are serviced according to their contractual terms. Otherwise, Assumption 21 (consequences of cover pool default) below applies.

(b) All outstanding other liabilities and the bank's liquidity line (if applicable) become due and payable immediately.

(c) The bank's other assets and pledged strategic cover pool assets are liquidated.

(d) The priority of payments is as follows:

- *The creditors of other liabilities only have a claim against the general insolvency estate, which primarily consists of the liquidation proceeds of non-pledged other assets.*

- *The bank's liquidity line has a priority claim on the liquidation proceeds of pledged strategic cover pool assets and pledged other assets and, if not sufficient, it also has a claim against the general insolvency estate. Liquidation proceeds of pledged assets which are not needed to repay the bank's liquidity line are released to the general insolvency estate.*

- *All claims against the general insolvency estate rank pari passu.*

- *Shareholders do not have a specific claim, they get whatever is left after all other claims have been fully satisfied.*

According to Assumption 19, a standalone cover pool under the administration of a cover pool administrator is created if the cover pool survives the event of bank default. This is one out of three potential options allowed for by the PfandBG, cf. Section 2.1, the other two being a cover pool transfer to another Pfandbrief bank and the administration of the cover pool by a cover pool administrator in a fiduciary capacity for another Pfandbrief bank. In all three cases, outstanding Pfandbriefe are serviced according to their contractual terms. As pointed out by Deutscher Bundestag [47], the most reasonable choice in practice would probably be to transfer the whole cover pool to another Pfandbrief bank. For modelling purposes this assumption is, however, not practicable as it needs strong additional assumptions regarding the composition of the resulting merged cover pool. In the following, we therefore assume that a standalone cover pool is created. The assumptions in (b)–(d) are also a reasonable approximation of the consequences of a bank default in practice: According to Koppmann, pp. 319–330, bank default triggers the bank's non-cover pool assets to be liquidated and the bank's debt (except for Pfandbriefe) to become due and payable immediately, irrespective of their contractual maturities. In our setup, the non-cover pool assets to be liquidated correspond to the bank's other assets and the pledged strategic cover pool assets. The priority of payments as specified by Assumption 19 is as in Merton [124], except for the ranking of the bank's liquidity line which is a consequence of our specific assumptions in the context of funding. It differs from the assumption made by Sünderhauf [148] who assumes that Pfandbrief holders also have a priority claim on the bank's other assets, while creditors of other liabilities only receive money once Pfandbrief holders have been fully repaid. As discussed in Section 3.3.2.1, Sünderhauf's approximation is not fully in line with the true Pfandbrief mechanics, but it is necessary as a dynamic reestablishment of cover requirements cannot be modelled in his one-period setup.

Figure 4.6 shows how the balance sheet changes upon bank default in case of a cover pool survival. Other assets and pledged strategic cover pool assets are liquidated and disappear, but non-pledged cover pool assets remain on the balance sheet. Other liabilities, the bank's liquidity line (if relevant) and the bank's equity also disappear as they are repaid as far as possible. Instead, new balance sheet positions are created: a cover pool liquidity line (if relevant) and subordinate residual claims. The latter can be raised again once all outstanding Pfandbriefe and the cover pool's liquidity line have been fully repaid.

4.5.2 Cover Pool Default

The bank is responsible for maintaining cover requirements and making contractual payments as long as it has not defaulted. Cover pool default therefore only becomes relevant upon bank default. As opposed to other covered bond frameworks, the PfandBG does not specify concrete default triggering events for the cover pool, cf. Section 2.1. This is why, in the following, we define cover pool default in a similar way to bank default.

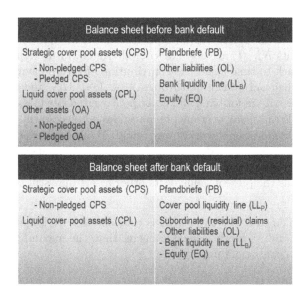

Figure 4.6: Balance sheet changes upon bank default in case the cover pool survives.

Definition 4.27 (Cover pool overindebtedness)
The cover pool is said to be overindebted at time $t \in \mathcal{T}$, $\tau \leq t \leq \tau^*$, if

$$V_P^O(t) < B_P^O(t), \tag{OP}$$

with

$$V_P^O(t) := V_{NPL}^{CPS}(t) + N^{CPL}(t),$$
$$B_P^O(t) := B_{PB}^O(t) + N_P^{LL}(t),$$
$$V_{NPL}^{CPS}(t) := \sum_{i=1}^{n_{CPS}} N_{i,NPL}^{CPS}(t) \cdot V_{i,U}^{CPS}(t),$$

and $B_{PB}^O(t)$ as in Definition 4.25.

The criterion for cover pool overindebtedness as defined in (OP) compares the value $V_P^O(t)$ of non-pledged cover pool assets to the exogenously given solvency barrier $B_P^O(t)$, which accounts for outstanding Pfandbrief nominals at time t and includes Pfandbrief repayments due at that time. Liquidity line repayments $N_P^{LL}(t)$ are also considered. Again, future repayments are discounted back to time t and the contribution of future payments further depends on their actual timing through T_{ST}, T_{LT}, M_{ST} and M_{LT}. When comparing the criteria for bank and cover pool overindebtedness in (OB) and (OP), one finds that bank overindebtedness does not necessarily trigger cover pool overindebtedness or the other way round. While bank overindebtedness could be caused by a sole decline in the value of the bank's other assets, cover pool overindebtedness does not necessarily imply that the bank is also overindebted as the present value of its other assets may still be sufficiently high.

As in the bank's case, the cover pool is considered to be illiquid if it is not overindebted and its funding need is bigger than the maximum amount of funding which can be obtained.

Definition 4.28 (Cover pool illiquidity)
The cover pool is said to be illiquid at time $t \in \mathcal{T}$, $\tau \le t \le \tau^*$, if

$$V_P^O(t) \ge B_P^O(t) \quad \text{and} \quad G_P^C(t) > L_P^{\max}(t). \tag{IP}$$

The specification of the event of cover pool default is now straightforward.

Assumption 20 (Cover pool default)
The time τ^* of cover pool default is given by

$$\tau^* := \inf \{t \in \mathcal{T}, \ t \ge \tau \mid (OP) \text{ or } (IP) \text{ holds}\},$$

with $\inf \emptyset := \infty$ *and* $\tau^* = \infty$ *being interpreted as no cover pool default occurring until time* $T_{\max}$.

As (OP) and (IP) cannot hold at the same time, there is always a well-defined reason for cover pool default, see Figure 4.7.

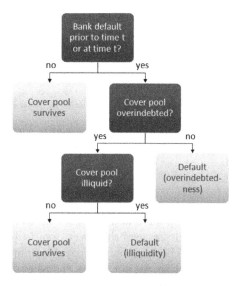

Figure 4.7: Cover pool default reasons at time $t \leq \tau^*$.

Regarding the consequences of cover pool default, we make the following assumption:

Assumption 21 (Consequences of cover pool default)
The consequences of cover pool default at time $\tau^* \in \mathcal{T}$ are as follows:

(a) All outstanding Pfandbriefe and the cover pool's liquidity line (if applicable) become due and payable immediately.

(b) All non-pledged cover pool assets are liquidated.

(c) The priority of payments is as follows:

- *Pfandbrief holders and the cover pool's liquidity line have a priority claim on the liquidation proceeds of non-pledged cover pool assets. Their claims all rank pari passu. If the liquidation proceeds of non-pledged cover pool assets are not sufficient, these creditors also have a claim against the general insolvency estate but only in case of a simultaneous bank default. Liquidation proceeds of non-pledged cover pool assets which are not needed to repay these creditors are released to the general insolvency estate.*

- *In case the bank had already defaulted at some previous time $\tau < \tau^*$,
  the residual claims of other liabilities and the bank's liquidity line
  which could not be satisfied during the insolvency proceedings at
  time τ can be raised again but only against the general insolvency
  estate.*

- *All claims against the general insolvency estate rank pari passu.*

- *Shareholders do not have a specific claim, they get whatever is left
  after all other claims (including residual claims) have been fully
  satisfied.*

Assumption 21 is a reasonable approximation of what could happen in
practice in case of a cover pool default. According to Koppmann [102], pp.
350–360, cover pool default triggers the cover pool to be liquidated and all
outstanding Pfandbriefe to become due and payable immediately, irrespective
of their contractual maturities. In our setup, the assets which remain in the
cover pool at the time of cover pool default correspond to the non-pledged
cover pool assets. The assumption that liquidation proceeds which are not
needed to satisfy the priority claim of Pfandbrief holders are released to
the bank's general insolvency estate is derived from the PfandBG.[15] The
priority of payments as specified by Assumption 21 reflects the Pfandbrief's
dual nature of protection, i.e. the priority claim on cover pool liquidation
proceeds and the additional unsecured claim against the general insolvency
estate. According to Assumption 21, this unsecured claim can only be raised
if the bank and the cover pool default at the same time. This is motivated
by the argument that in practice it might be hard to enforce such a claim if
bank and cover pool insolvency proceedings take place at different points
in time.[16] The ranking of the cover pool's liquidity line is a consequence of
Assumption 15.

Before we start with the specification of the liquidation payments, we introduce some more notation which will be needed to distinguish the nine
Pfandbrief scenarios as described in Section 4.1 above. With the help of
these scenarios, we are able to incorporate all required modelling features

---

[15]The PfandBG [157], § 30, par. 4, sent. 2, states that "[a]ssets remaining after the
Pfandbrief creditors are satisfied [...] must be surrendered to the insolvent estate".

[16]In their recovery analysis, Fitch completely ignores the Pfandbrief holder's recourse to
the general insolvency estate by referring to the potential impracticality to enforce
such a claim if the bank and the cover pool insolvency proceeding do not start at the
same time, cf. Muños and Mezza [129].

as found in Section 2.3 into our model, especially the Pfandbrief mechanics and potential asset liability management activities.

Pfandbrief scenario at time t The Pfandbrief scenario at time t is denoted by

$$S(t) := \begin{cases} 1, & \text{if } t < \tau \leq \tau^* \text{ and } t < T_{\max}, \\ 2, & \text{if } t = \tau < \tau^* \text{ and } t < T_{\max}, \\ 3, & \text{if } t = \tau = \tau^* \text{ and } t < T_{\max}, \\ 4, & \text{if } \tau < t < \tau^* \text{ and } t < T_{\max}, \\ 5, & \text{if } \tau < t = \tau^* \text{ and } t < T_{\max}, \\ 6, & \text{if } \tau \leq \tau^* < t \text{ and } t < T_{\max}, \\ 7, & \text{if } t \leq \tau \leq \tau^* \text{ and } t = T_{\max}, \\ 8, & \text{if } \tau < t \leq \tau^* \text{ and } t = T_{\max}, \\ 9, & \text{if } \tau \leq \tau^* < t \text{ and } t = T_{\max}. \end{cases}$$

Obviously, $\cup_{i=1,\dots,9} \{S(t) = i\} = \Omega$ and $\{S(t) = i\} \cap \{S(t) = j\} = \emptyset$ for $i \neq j$, $i, j = 1, \dots, 9$.

4.5.3 Liquidation Payments

Having defined the events of bank and cover pool default, we now specify the liquidation payments made to creditors at time $t \in \{\tau, \tau^*, T_{\max}\}$, $\tau, \tau^* \in \mathcal{T}$. These payments depend on the realized liquidation proceeds, the creditors' claims and the priority of payments. We distinguish forced asset liquidations triggered by an event of default at some time prior to $T_{\max}$ from planned asset liquidations in $T_{\max}$, which arise irrespective of an event of default. Regarding the prices at which assets can be liquidated, we make the following assumption:

Assumption 22 (Liquidation haircuts)
With respect to asset liquidations, the following holds:

(a) In case of a forced asset liquidation due to an event of bank or cover pool default at time $t \in \mathcal{T}$, $t < T_{\max}$, risky assets can only be liquidated at a haircut to their present value. This haircut is $h_{LI}^{OA} \in [0, 1]$ for other assets and $h_{LI}^{CPS} \in [0, 1]$ for strategic cover pool assets. No liquidation

haircut applies for liquid cover pool assets and realized time-t asset cash inflows.

(b) In case of a planned asset liquidation in $T_{\max}$, all asset can be sold at their full present value.

The assumption that asset sales which are triggered by an event of bank or cover pool default at some time prior to $T_{\max}$ are only possible at a haircut to the assets' present value is in line with our previous discussions in Section 3.3.1. If assets need to be sold on short notice this may not be possible at or close to their present value, especially in times of market distress. As liquid cover pool assets and realized time-t asset cash inflows already constitute risk-free cash positions, no liquidation haircuts apply in their case. Assuming a liquidation haircut of zero in $T_{\max}$ is motivated by the argument that planned asset liquidations do not come as a surprise, meaning that there is sufficient time for the bank or the cover pool administrator to prepare these asset sales.

Observation 4.29 (Asset liquidation proceeds)
At time $t \in \mathcal{T}$, the proceeds from asset liquidation are given by

$$LV_{PL}^{OA}(t) := \begin{cases} \left(1 - h_{LI}^{OA}\right) \cdot \tilde{V}_{PL}^{OA}(t) + \tilde{F}_{PL}^{OA}(t), & \text{if } S(t) \in \{2, 3\}, \\ V_{PL}^{OA}(t), & \text{if } S(t) = 7, \\ 0, & \text{otherwise}, \end{cases}$$

$$LV_{NPL}^{OA}(t) := \begin{cases} \left(1 - h_{LI}^{OA}\right) \cdot \tilde{V}_{NPL}^{OA}(t) + \tilde{F}_{NPL}^{OA}(t), & \text{if } S(t) \in \{2, 3\}, \\ V_{NPL}^{OA}(t), & \text{if } S(t) = 7, \\ 0, & \text{otherwise}, \end{cases}$$

$$LV_{PL}^{CP}(t) := \begin{cases} \left(1 - h_{LI}^{CPS}\right) \cdot \tilde{V}_{PL}^{CPS}(t) + \tilde{F}_{PL}^{CPS}(t), & \text{if } S(t) \in \{2, 3\}, \\ V_{PL}^{CPS}(t), & \text{if } S(t) = 7, \\ 0, & \text{otherwise}, \end{cases}$$

$$LV_{NPL}^{CP}(t) := \begin{cases} \left(1 - h_{LI}^{CPS}\right) \cdot \tilde{V}_{NPL}^{CPS}(t) \\ \qquad + \tilde{F}_{NPL}^{CPS}(t) + N^{CPL}(t), & \text{if } S(t) \in \{3, 5\}, \\ V_{NPL}^{CPS}(t) + N^{CPL}(t), & \text{if } S(t) \in \{7, 8\}, \\ 0, & \text{otherwise}, \end{cases}$$

with $\tilde{V}^x_y(t) := \sum_{i \in \mathcal{N}_x, T^x_i > t} N^x_{i,y}(t) \cdot V^x_{i,u}(t)$ and $\tilde{F}^x_y(t) := \sum_{i=1}^{n_x} N^x_{i,y}(t) \cdot \tilde{F}^x_{i,u}(t)$ for $x \in \{CPS, OA\}$ and $y \in \{PL, NPL\}$. The bank's total liquidation proceeds are given by

$$LV(t) := LV^{OA}(t) + LV^{CP}(t),$$

where $LV^z(t) := LV^z_{PL}(t) + LV^z_{NPL}(t)$, $z \in \{OA, CP\}$.

Proof. The statement is a consequence of Assumptions 3, 19, 21 and 22 and takes into consideration that positions in pledged assets are 0 in $t > \tau$, cf. Assumptions 13, 15 and 19.   □

The literature distinguishes three different ways of measuring recovery payments in case of default, see Lando [108], pp. 120–122: recovery of face value, recovery of treasury value and recovery of market value. The *recovery of face value* assumes that debt of the same priority is assigned a fractional recovery based on the outstanding nominal, ignoring different maturities and coupons. According to Lando, this is the closest to legal practice out of the three measures, and it is the measure typically used by rating agencies. Under the *recovery of treasury value* assumption, the defaulted bond is replaced by a risk-free bond with the same maturity but with a reduced nominal, which is better in line with the economic perspective that nominals of bonds with longer maturities should be discounted more than those of bonds with shorter maturities. The *recovery of market value* measures the change in market value at the time of default, i.e. the market value loss associated with the default event, which has an economic meaning. For zero coupon bonds which can only default at maturity (i.e. the risky assets in our model) all concepts coincide. Pfandbriefe and other liabilities, on the other hand, can also default prior to maturity in our model. In the following, we consider nominal repayments to these liabilities to be due at maturity rather than at the time of default (similar to the recovery of treasury value assumption) and calculate the corresponding claims by discounting the outstanding nominals back to the time of default. For the liquidity lines, there is no need for discounting as the corresponding payments are due at the time of default. In case of planned asset liquidations in $T_{\max}$, the same concept as for forced asset liquidation applies.

Assumption 23 (Liquidation claims)
At time $t \in \mathcal{T}$, the creditors' liquidation claims are given by

$$C_k^{PB}(t) := \begin{cases} N_k^{PB}(t) \cdot d(t, T_k^{PB}), & \text{if } S(t) \in \{3,5,7,8\}, \\ 0, & \text{otherwise}, \end{cases}$$

$$C_l^{OL}(t) := \begin{cases} N_l^{OL}(t) \cdot d(t, T_l^{OL}), & \text{if } S(t) \in \{2,3,7\}, \\ RC_l^{OL}(\tau), & \text{if } S(t) \in \{5,8\}, \\ 0, & \text{otherwise}, \end{cases}$$

$$C_B^{LL}(t) := \begin{cases} N_B^{LL}(t), & \text{if } S(t) \in \{2,3,7\}, \\ RC_B^{LL}(\tau), & \text{if } S(t) \in \{5,8\}, \\ 0, & \text{otherwise}, \end{cases}$$

$$C_P^{LL}(t) := \begin{cases} N_P^{LL}(t), & \text{if } S(t) \in \{5,8\}, \\ 0, & \text{otherwise}, \end{cases}$$

for $k = 1, ..., n_{PB}$ and $l = 1, ..., n_{OL}$, where $d(t, T)$ is as in Definition 4.25. For $S(t) \in \{5,8\}$, the residual claims from an earlier bank default at time $\tau < \min(\tau^*, T_{\max})$ are

$$RC_l^{OL}(\tau) := \max(0; C_l^{OL}(\tau) - \hat{F}_l^{OL}(\tau)),$$
$$RC_B^{LL}(\tau) := \max(0; C_B^{LL}(\tau) - \hat{F}_B^{LL}(\tau)),$$

with $\hat{F}_l^{OL}(\tau)$ and $\hat{F}_B^{LL}(\tau)$ being the realized liquidation payments from the time $\tau < \min(\tau^*, T_{\max})$ of bank default as in Observation 4.30 below. The total claims of Pfandbrief holders and creditors of other liabilities are given by

$$C^{PB}(t) := \sum_{k=1}^{n_{PB}} C_k^{PB}(t) \quad \text{and} \quad C^{OL}(t) := \sum_{l=1}^{n_{OL}} C_l^{OL}(t).$$

Equity does not have a specific claim but gets the remainder after all other liabilities have been repaid.

Regarding the priority of payments in $T_{\max}$, we make the following assumption:

Assumption 24 (Priority of payments in $T_{\max}$)
At time $T_{\max}$, the priority of payments is as follows:

- *If the bank has not defaulted prior to time $T_{\max}$, the creditors of other liabilities have a claim against the general liquidation estate, which primarily consists of the liquidation proceeds of non-pledged other assets. Furthermore, the bank's liquidity line has a priority claim on the liquidation proceeds of pledged assets and, if not sufficient, it also has a claim against the general liquidation estate. Liquidation proceeds of pledged assets which are not needed to repay the bank's liquidity line are released to the general liquidation estate.*

- *If the cover pool has not defaulted prior to time $T_{\max}$, Pfandbrief holders and the cover pool's liquidity line have a priority claim on the liquidation proceeds of non-pledged cover pool assets, and their claims all rank pari passu. If the liquidation proceeds of non-pledged cover pool assets are not sufficient to satisfy these claims, Pfandbrief holders and the cover pool's liquidity line also have a claim against the general liquidation estate. Liquidation proceeds of non-pledged cover pool assets which are not needed to repay these creditors are released to the general liquidation estate.*

- *If the bank has already defaulted at some time $\tau < T_{\max}$ but the cover pool has not, the residual claims of other liabilities and the bank's liquidity line which could not be satisfied during the bank's insolvency proceedings can be raised again but only against the general liquidation estate.*

- *All claims against the general liquidation estate rank pari passu.*

- *Shareholders do not have a specific claim, they get whatever is left after all other claims (including residual claims from an earlier time $\tau < T_{\max}$ of bank default) have been fully satisfied.*

In case the bank has not defaulted prior to $T_{\max}$, the priority of payments as specified above corresponds to the one in case of a simultaneous bank and cover pool default, cf. Assumptions 19 and 21. Similarly, if the bank has defaulted prior to $T_{\max}$ but the cover pool has not, the priority of payments is the same as in the case of a standalone cover pool default at some time $t > \tau$.

Having specified the realized liquidation proceeds, the creditors' claims and the priority of payments in case of a liquidation event, the liquidation payments to creditors can now be derived as follows.

Observation 4.30 (Liquidation payments)
At time $t \in \mathcal{T}$, the liquidation payments to creditors are given by

$$\hat{F}^{PB}(t) := \min\left(C^{PB}(t); k_{PB}^{CP}(t) \cdot LV_{NPL}^{CP}(t) + k_{PB}^{G}(t) \cdot LV^{IE}(t)\right),$$

$$\hat{F}^{OL}(t) := \min\left(C^{OL}(t); k_{OL}^{G}(t) \cdot LV^{IE}(t)\right),$$

$$\hat{F}_{B}^{LL}(t) := \min\left(C_{B}^{LL}(t); k_{LB}^{PA}(t) \cdot LV_{PL}^{A}(t) + k_{LB}^{G}(t) \cdot LV^{IE}(t)\right),$$

$$\hat{F}_{P}^{LL}(t) := \min\left(C_{P}^{LL}(t); k_{LP}^{CP}(t) \cdot LV_{NPL}^{CP}(t) + k_{LP}^{G}(t) \cdot LV^{IE}(t)\right),$$

$$\hat{F}^{EQ}(t) := \max\left(0; LV(t) - C^{PB}(t) - C^{OL}(t) - C_{B}^{LL}(t) - C_{P}^{LL}(t)\right),$$

with $LV_{PL}^{A}(t) := LV_{PL}^{OA}(t) + LV_{PL}^{CP}(t)$ *and*

$$k_{PB}^{CP}(t) := \begin{cases} \frac{C^{PB}(t)}{C^{PB}(t)+C_{P}^{LL}(t)}, & \text{if } C^{PB}(t) + C_{P}^{LL}(t) > 0, \\ 0, & \text{otherwise,} \end{cases}$$

$$k_{LP}^{CP}(t) := \begin{cases} \frac{C_{P}^{LL}(t)}{C^{PB}(t)+C_{P}^{LL}(t)}, & \text{if } C^{PB}(t) + C_{P}^{LL}(t) > 0, \\ 0, & \text{otherwise,} \end{cases}$$

$$k_{LB}^{PA}(t) := \begin{cases} 1, & \text{if } C_{B}^{LL}(t) > 0, \\ 0, & \text{otherwise,} \end{cases}$$

$$k_{PB}^{G}(t) := \begin{cases} \frac{G^{PB}(t)}{C^{IE}(t)}, & \text{if } C^{IE}(t) > 0, \\ 0, & \text{otherwise,} \end{cases}$$

$$k_{OL}^{G}(t) := \begin{cases} \frac{C^{OL}(t)}{C^{IE}(t)}, & \text{if } C^{IE}(t) > 0, \\ 0, & \text{otherwise,} \end{cases}$$

$$k_{LB}^{G}(t) := \begin{cases} \frac{G_{B}^{LL}(t)}{C^{IE}(t)}, & \text{if } C^{IE}(t) > 0, \\ 0, & \text{otherwise,} \end{cases}$$

$$k_{LP}^{G}(t) := \begin{cases} \frac{G_{P}^{LL}(t)}{C^{IE}(t)}, & \text{if } C^{IE}(t) > 0, \\ 0, & \text{otherwise,} \end{cases}$$

where

$$LV^{IE}(t) := LV_{NPL}^{OA}(t) + E^{PA}(t) + E^{CP}(t),$$

$$E^{PA}(t) := \max\left(0; LV_{PL}^{OA}(t) + LV_{PL}^{CP}(t) - C_B^{LL}(t)\right),$$

$$E^{CP}(t) := \max\left(0; LV_{NPL}^{CP}(t) - C^{PB}(t) - C_P^{LL}(t)\right),$$

$$C^{IE}(t) := C^{OL}(t) + G^{PB}(t) + G_B^{LL}(t) + G_P^{LL}(t),$$

$$G_B^{LL}(t) := \max\left(0; C_B^{LL}(t) - k_{LB}^{PA}(t) \cdot LV_{PL}^A(t)\right),$$

$$G^{PB}(t) := \max\left(0; C^{PB}(t) - k_{PB}^{CP}(t) \cdot LV_{NPL}^{CP}(t)\right),$$

$$G_P^{LL}(t) := \max\left(0; C_P^{LL}(t) - k_{LP}^{CP}(t) \cdot LV_{NPL}^{CP}(t)\right).$$

For single Pfandbriefe and other liabilities, the liquidation payments are given by

$$\hat{F}_k^{PB}(t) := \begin{cases} \frac{\hat{F}^{PB}(t)}{C^{PB}(t)} \cdot C_k^{PB}(t), & \text{if } C^{PB}(t) > 0, \\ 0, & \text{otherwise,} \end{cases}$$

$$\hat{F}_l^{OL}(t) := \begin{cases} \frac{\hat{F}^{OL}(t)}{C^{OL}(t)} \cdot C_l^{OL}(t), & \text{if } C^{OL}(t) > 0, \\ 0, & \text{otherwise,} \end{cases}$$

for $k = 1, ..., n_{PB}$ and $l = 1, ..., n_{OL}$.

Proof. According to the above definitions of $\hat{F}^{PB}(t)$, $\hat{F}^{OL}(t)$, $\hat{F}_B^{LL}(t)$, $\hat{F}_P^{LL}(t)$ and $\hat{F}^{EQ}(t)$, non-zero liquidation payments only arise in case of non-zero claims, i.e. for $S(t) \in \{2, 3, 5, 7, 8\}$. The statement then follows from the priority of payments a specified by Assumptions 19, 21 and 24: Pfandbrief holders and the cover pool's liquidity line have a priority claim on the liquidation proceeds of non-pledged cover pool assets $LV_{NPL}^{CPS}(t)$ and the bank's liquidity line has a priority claim on the liquidation proceeds of pledged assets $LV_{PL}^{CPS}(t)$ and $LV_{PL}^{OA}(t)$. Liquidation proceeds of non-pledged other assets $LV_{NPL}^{OA}(t)$ and liquidation proceeds which are not needed to satisfy the priority claims, i.e. $E^{CP}(t)$ and $E^{PA}(t)$, belong to the general insolvency estate $LV^{IE}(t)$. Furthermore, the fractions $k_{PB}^{CP}(t)$ and $k_{LP}^{CP}(t)$ represent the priority claim of Pfandbrief holders and the cover pool's liquidity line on the cover pool and $k_{LB}^{PA}(t)$ reflects the priority claim of the bank's liquidity line on the pledged assets. The fractions $k_{PB}^G(t)$, $k_{OL}^G(t)$, $k_{LB}^G(t)$ and $k_{LP}^G(t)$, on the other hand, represent the unsecured claims of Pfandbrief holders, the creditors of other liabilities, the bank's liquidity line and the cover pool's liquidity line against $LV^{IE}(t)$. The formulas for single Pfandbriefe and other liabilities follow from the fact that, within each of the two liability classes, claims rank pari passu. $\qquad\qquad\qquad\qquad\qquad\qquad\qquad\qquad\qquad\qquad\quad$ □

Remark 4.31 (Liquidation payment to the cover pool's liquidity line)
At time $t \in \mathcal{T}$, the liquidation payment to the cover pool's liquidity line can be written as

$$\hat{F}_P^{LL}(t) = \min\left(C_P^{LL}(t); k_{LP}^{CP}(t) \cdot LV_{NPL}^{CP}(t)\right).$$

Proof. It is sufficient to show that $k_{LP}^{G}(t) \cdot LV^{IE}(t)$ is always 0. In the case where $S(t) \notin \{5,8\}$ we have $C_P^{LL}(t) = 0$ and therefore $k_{LP}^{G}(t) = 0$. For $S(t) \in \{5,8\}$ we distinguish two sub cases: $G_P^{LL}(t) = 0$ and $G_P^{LL}(t) > 0$. If $G_P^{LL}(t) = 0$ we get $k_{LP}^{G}(t) = 0$. For $G_P^{LL}(t) > 0$ we have

$$C_P^{LL}(t) > k_{LP}^{CP}(t) \cdot LV_{NPL}^{CP}(t) = \frac{C_P^{LL}(t)}{C^{PB}(t) + C_P^{LL}(t)} \cdot LV_{NPL}^{CP}(t),$$

which implies
$$C^{PB}(t) + C_P^{LL}(t) > LV_{NPL}^{CP}(t),$$
and with $LV^{OA}(t) = E^{PA}(t) = 0$ it follows that

$$LV^{IE}(t) = E^{CP}(t) = \max\left(0; LV_{NPL}^{CP}(t) - C^{PB}(t) - C_P^{LL}(t)\right) = 0.$$

$\qquad\qquad\qquad\qquad\qquad\qquad\qquad\qquad\qquad\qquad\qquad\qquad\qquad\qquad$ □

The liquidation payments in Observation 4.30 always add up to the total liquidation proceeds, i.e. the latter are well-specified:

Proposition 4.32 (Liquidation payments)
At time $t \in \mathcal{T}$, it holds that

$$\hat{F}^{PB}(t) + \hat{F}^{OL}(t) + \hat{F}_B^{LL}(t) + \hat{F}_P^{LL}(t) + \hat{F}^{EQ}(t) = LV(t). \qquad (4.26)$$

Proof. To prove the statement, we distinguish four cases:

Case 1: $E^{CP}(t) > 0$ and $E^{PA}(t) > 0$. In this case we have

$$G^{PB}(t) = G_P^{LL}(t) = G_B^{LL}(t) = 0,$$

and with $k_{OL}^G(t) = 1$ and $k_{PB}^G(t) = k_{LB}^G(t) = k_{LP}^G(t) = 0$ we get

$$\hat{F}^{PB}(t) = \min\left(C^{PB}(t); k_{PB}^{CP}(t) \cdot LV_{NPL}^{CP}(t)\right) = C^{PB}(t),$$
$$\hat{F}_P^{LL}(t) = \min\left(C_P^{LP}(t); k_{LP}^{CP}(t) \cdot LV_{NPL}^{CP}(t)\right) = C_P^{LL}(t),$$
$$\hat{F}_B^{LL}(t) = \min\left(C_B^{LL}(t); k_{LB}^{PA}(t) \cdot \left(LV_{PL}^{OA}(t) + LV_{PL}^{CP}(t)\right)\right) = C_B^{LL}(t).$$

Furthermore, it holds that

$$LV^{IE}(t) = LV(t) - C^{PB}(t) - C_B^{LL}(t) - C_P^{LL}(t),$$

and therefore

$$\hat{F}^{OL}(t) = \min\left(C^{OL}(t); LV(t) - C^{PB}(t) - C_B^{LL}(t) - C_P^{LL}(t)\right),$$
$$\hat{F}^{EQ}(t) = \max\left(0; LV(t) - C^{PB}(t) - C^{OL}(t) - C_B^{LL}(t) - C_P^{LL}(t)\right).$$

Summation yields (4.26).

Case 2: $E^{CP}(t) > 0$ and $E^{PA}(t) = 0$. Here, we have

$$G^{PB}(t) = G_P^{LL}(t) = 0,$$

and with $k_{PB}^G(t) = k_{LB}^G(t) = 0$ it follows that

$$\hat{F}^{PB}(t) = \min\left(C^{PB}(t); k_{PB}^{CP}(t) \cdot LV_{NPL}^{CP}(t)\right) = C^{PB}(t),$$
$$\hat{F}_P^{LL}(t) = \min\left(C_P^{LL}(t); k_{LP}^{CP}(t) \cdot LV_{NPL}^{CP}(t)\right) = C_P^{LL}(t).$$

It also holds that

$$LV^{IE}(t) = LV_{NPL}^{OA}(t) + LV_{NPL}^{CP}(t) - C^{PB}(t) - C_P^{LL}(t),$$
$$C^{IE}(t) = C^{OL}(t) + G_B^{LL}(t).$$

We now distinguish two sub cases: For $LV^{IE}(t) < C^{IE}(t)$, we get

$$\hat{F}^{OL}(t) = k_{OL}^G(t) \cdot LV^{IE}(t),$$

$$\hat{F}_B^{LL}(t) = k_{LB}^{PA}(t) \cdot \left(LV_{PL}^{OA}(t) + LV_{PL}^{CP}(t)\right) + k_{LB}^G(t) \cdot LV^{IE}(t),$$
$$\hat{F}^{EQ}(t) = 0,$$

and, with $k_{OL}^G(t) + k_{LB}^G(t) = 1$,

$$\hat{F}^{OL}(t) + \hat{F}_B^{LL}(t) = LV^{IE}(t) + k_{LB}^{PA}(t) \cdot \left(LV_{PL}^{OA}(t) + LV_{PL}^{CP}(t)\right),$$
$$= LV^{IE}(t) + LV_{PL}^{OA}(t) + LV_{PL}^{CP}(t),$$
$$= LV(t) - C^{PB}(t) - C_P^{LL}(t).$$

Note that the second equality holds because of $k_{LB}^{PA}(t) \in \{0,1\}$. For $k_{LB}^{PA}(t) = 1$ the statement is obvious and $k_{LB}^{PA}(t) = 0$ implies $C_B^{LL}(t) = 0$, which together with $E^{PA}(t) = 0$ means that $LV_{PL}^{OA}(t) = LV_{PL}^{CP}(t) = 0$.

On the other hand, if $LV^{IE}(t) \geq C^{IE}(t)$ it follows that

$$\hat{F}^{OL}(t) = C^{OL}(t),$$
$$\hat{F}_B^{LL}(t) = C_B^{LL}(t),$$
$$\hat{F}^{EQ}(t) = LV(t) - C^{PB}(t) - C^{OL}(t) - C_B^{LL}(t) - C_P^{LL}(t).$$

In both cases summation yields (4.26).

Case 3: $E^{CP}(t) = 0$ and $E^{PA}(t) > 0$. In this case we have $G_B^{LL}(t) = 0$ and, with $k_{LB}^G(t) = 0$,

$$\hat{F}_B^{LL}(t) = \min\left(C_B^{LL}(t); k_{LB}^{PA}(t) \cdot \left(LV_{PL}^{OA}(t) + LV_{PL}^{CP}(t)\right)\right) = C_B^{LL}(t).$$

In addition,

$$LV^{IE}(t) = LV^{OA}(t) + LV_{PL}^{CP}(t) - C_B^{LL}(t),$$
$$C^{IE}(t) = C^{OL}(t) + G^{PB}(t) + G_P^{LL}(t).$$

For $LV^{IE}(t) < C^{IE}(t)$ this results in

$$\hat{F}^{PB}(t) = k_{PB}^{CP}(t) \cdot LV_{NPL}^{CP}(t) + k_{PB}^G(t) \cdot LV^{IE}(t),$$
$$\hat{F}^{OL}(t) = k_{OL}^G(t) \cdot LV^{IE}(t),$$
$$\hat{F}_P^{LL}(t) = k_{LP}^{CP}(t) \cdot LV_{NPL}^{CP}(t) + k_{LP}^G(t) \cdot LV^{IE}(t),$$
$$\hat{F}^{EQ}(t) = 0,$$

and with $k_{PB}^{CP}(t) + k_{LP}^{CP}(t) = 1$ and $k_{PB}^G(t) + k_{OL}^G(t) + k_{LP}^G(t) = 1$ we get

$$\hat{F}^{PB}(t) + \hat{F}^{OL}(t) + \hat{F}_P^{LL}(t) = LV_{NPL}^{CP}(t) + LV^{IE}(t) = LV(t) - C_B^{LL}(t).$$

For $LV^{IE}(t) \geq C^{IE}(t)$, we have

$$\hat{F}^{PB}(t) = C^{PB}(t),$$
$$\hat{F}^{OL}(t) = C^{OL}(t),$$
$$\hat{F}_P^{LL}(t) = C_P^{LL}(t),$$
$$\hat{F}^{EQ}(t) = LV(t) - C^{PB}(t) - C^{OL}(t) - C_B^{LL}(t) - C_P^{LL}(t).$$

Again, in both cases summation yields (4.26).

Case 4: $E^{CP}(t) = 0$ and $E^{PA}(t) = 0$. It then holds that

$$LV^{IE}(t) = LV_{NPL}^{OA}(t),$$
$$C^{IE}(t) = C^{OL}(t) + G^{PB}(t) + G_B^{LL}(t) + G_P^{LL}(t).$$

For $LV^{IE}(t) < C^{IE}(t)$ it follows that

$$\hat{F}^{PB}(t) = k_{PB}^{CP}(t) \cdot LV_{NPL}^{CP}(t) + k_{PB}^G(t) \cdot LV^{IE}(t),$$
$$\hat{F}^{OL}(t) = k_{OL}^G(t) \cdot LV^{IE}(t),$$
$$\hat{F}_B^{LL}(t) = k_{LB}^{PA}(t) \cdot \left(LV_{PL}^{OA}(t) + LV_{PL}^{CP}(t) \right) + k_{LB}^G(t) \cdot LV^{IE}(t),$$
$$\hat{F}_P^{LL}(t) = k_{LP}^{CP}(t) \cdot LV_{NPL}^{CP}(t) + k_{LP}^G(t) \cdot LV^{IE}(t),$$
$$\hat{F}^{EQ}(t) = 0,$$

and with $k_{PB}^{CP}(t) + k_{LL}^{CP}(t) = 1$ and $k_{PB}^G(t) + k_{OL}^G(t) + k_{LB}^G(t) + k_{LP}^G(t) = 1$ this results in

$$\hat{F}^{PB}(t) + \hat{F}^{OL}(t) + \hat{F}_B^{LL}(t) + \hat{F}_P^{LL}(t)$$
$$= LV_{NPL}^{CP}(t) + LV^{IE}(t) + k_{LB}^{PA}(t) \cdot \left(LV_{PL}^{OA}(t) + LV_{PL}^{CP}(t) \right)$$
$$= LV_{NPL}^{CP}(t) + LV^{IE}(t) + LV_{PL}^{OA}(t) + LV_{PL}^{CP}(t)$$
$$= LV(t).$$

Again, the second equality holds because of $k_{LB}^{PA}(t) \in \{0, 1\}$.

On the other hand, if $LV^{IE}(t) \geq C^{IE}(t)$ we have

$$\hat{F}^{PB}(t) = C^{PB}(t),$$
$$\hat{F}^{OL}(t) = C^{OL}(t),$$
$$\hat{F}_B^{LL}(t) = C_B^{LL}(t),$$
$$\hat{F}_P^{LL}(t) = C_P^{LL}(t),$$
$$\hat{F}^{EQ}(t) = LV(t) - C^{PB}(t) - C^{OL}(t) - C_B^{LL}(t) - C_P^{LL}(t).$$

Summation then yields (4.26). $\square$

In case of overindebtedness, there is always a loss for the affected creditors.

Observation 4.33 (Default due to overindebtedness)
The followings holds in case of a default due to overindebtedness:

(a) If the bank is overindebted at time $\tau \in \mathcal{T}$, $\tau < \tau^*$, outstanding other liabilities and the bank's liquidity line (if drawn) incur a loss, i.e.

$$\hat{F}_l^{OL}(\tau) < C_l^{OL}(\tau) \quad \text{if } C_l^{OL}(\tau) > 0, \qquad l = 1, ..., n_{OL},$$
$$\hat{F}_B^{LL}(t) < C_B^{LL}(t) \quad \text{if } C_B^{LL}(t) > 0.$$

(b) If the cover pool is overindebted at time $\tau^* \in \mathcal{T}$, $\tau^* > \tau$, outstanding Pfandbriefe and the cover pool's liquidity line (if drawn) incur a loss, i.e.

$$\hat{F}_k^{PB}(\tau^*) < C_k^{PB}(\tau^*) \quad \text{if } C_k^{PB}(\tau^*) > 0, \qquad k = 1, ..., n_{PB},$$
$$\hat{F}_P^{LL}(\tau^*) < C_P^{LL}(\tau^*) \quad \text{if } C_P^{LL}(\tau^*) > 0.$$

(c) If both the bank and the cover pool are overindebted at time $\hat{\tau} \in \mathcal{T}$ with $\hat{\tau} = \tau = \tau^*$, outstanding Pfandbriefe and other liabilities as well as the bank's liquidity line (if drawn) incur a loss, i.e.

$$\hat{F}_k^{PB}(\hat{\tau}) < C_k^{PB}(\hat{\tau}) \quad \text{if } C_k^{PB}(\hat{\tau}) > 0, \qquad k = 1, ..., n_{PB},$$
$$\hat{F}_l^{OL}(\hat{\tau}) < C_l^{OL}(\hat{\tau}) \quad \text{if } C_l^{OL}(\hat{\tau}) > 0, \qquad l = 1, ..., n_{OL},$$
$$\hat{F}_B^{LL}(\hat{\tau}) < C_B^{LL}(\hat{\tau}) \quad \text{if } C_B^{LL}(\hat{\tau}) > 0.$$

The reverse statements are not true, i.e. it cannot be concluded that non-overindebtedness implies that there are no losses to the corresponding creditors. In case of a default due to illiquidity, liquidation proceeds might still not be sufficient to repay the corresponding creditors if the liquidation haircuts h_{LI}^{OA} and h_{LI}^{CPS} are positive. In addition, the default barriers $B_B^O(\tau)$ and $B_P^O(\tau^*)$ do not fully account for long term debt if $M_{LT} < 1$.

Proof. (a) If the bank defaults due to overindebtedness at time $\tau \in \mathcal{T}$, $\tau < \tau^*$, it holds that

$$V_{PL}^{CPS}(\tau) + V^{OA}(\tau) + V_{NPL}^{CPS}(\tau) + N^{CPL}(\tau) < B_{PB}^O(\tau) + B_{OL}^O(\tau) + N_B^{LL}(\tau),$$

and as the cover pool does not default at the same time, we have

$$V_{NPL}^{CPS}(\tau) + N^{CPL}(\tau) \geq B_{PB}^O(\tau).$$

It follows that

$$V_{PL}^{CPS}(\tau) + V^{OA}(\tau) < B_{OL}^O(\tau) + N_B^{LL}(\tau) + \underbrace{B_{PB}^O(\tau) - V_{NPL}^{CPS}(\tau) - N^{CPL}(\tau)}_{\leq 0}$$

$$< B_{OL}^O(\tau) + N_B^{LL}(\tau),$$

and, because of h_{LI}^{CPS}, $h_{LI}^{OA} \in [0,1]$ and $f_O(t,s) \leq 1 \ \forall s \geq t$, we get

$$LV_{PL}^{CP}(\tau) + LV^{OA}(\tau) \leq V_{PL}^{CPS}(\tau) + V^{OA}(\tau)$$
$$< B_{OL}^O(\tau) + N_B^{LL}(\tau)$$
$$\leq C^{OL}(\tau) + C_B^{LL}(\tau).$$

Furthermore, with

$$LV^{IE}(\tau) = LV_{NPL}^{OA}(\tau) + E^{PA}(\tau),$$
$$C^{IE}(\tau) = C^{OL}(\tau) + G_B^{LL}(\tau),$$

and

$$E^{PA}(\tau) - G_B^{LL}(\tau) = LV_{PL}^{OA}(\tau) + LV_{PL}^{CP}(\tau) - C_B^{LL}(\tau) \tag{4.27}$$

it follows that

$$LV^{IE}(\tau) - C^{IE}(\tau) = LV^{OA}(\tau) + LV_{PL}^{CP}(\tau) - C^{OL}(\tau) - C_B^{LL}(\tau) < 0.$$

Note that (4.27) holds as $k_{LB}^{PA}(\tau) \in \{0,1\}$. For $k_{LB}^{PA}(\tau) = 1$, the statement is obvious. On the other hand, $k_{LB}^{PA}(\tau) = 0$ implies $C_B^{LL}(\tau) = G_B^{LL}(\tau) = 0$ and therefore $E^{PA}(\tau) = LV_{PL}^{OA}(\tau) + LV_{PL}^{CP}(\tau)$. As $C^{OL}(\tau) > 0$, the liquidation payments to creditors of other liabilities are given by

$$\hat{F}^{OL}(\tau) = k_{OL}^G(\tau) \cdot LV^{IE}(\tau) < C^{OL}(\tau),$$

and for $C_B^{LL}(\tau) > 0$ we get

$$\hat{F}_B^{LL}(\tau) = LV_{PL}^{OA}(\tau) + LV_{PL}^{CP}(\tau) + k_{LB}^G(\tau) \cdot LV^{IE}(\tau) < C_B^{LL}(\tau).$$

The statement with respect to single other liabilities follows from the definition of $\hat{F}_l^{OL}(t)$ for $l = 1, ..., n_{OL}$.

(b) If the cover pool defaults due to overindebtedness at time $\tau^* \in \mathcal{T}$, $\tau^* > \tau$, it holds that

$$\begin{aligned} LV_{NPL}^{CP}(\tau^*) &\leq V_{NPL}^{CPS}(\tau^*) + N^{CPL}(\tau^*) \\ &< B_{PB}^O(\tau^*) + N_P^{LL}(\tau^*) \\ &\leq C^{PB}(\tau^*) + C_P^{LL}(\tau^*) \end{aligned}$$

as $h_{LI}^{CPS} \in [0,1]$ and $f_O(t,s) \leq 1 \ \forall s \geq t$. With $C^{PB}(\tau^*) > 0$, the liquidation payments to Pfandbrief holders are given by

$$\hat{F}^{PB}(\tau^*) = k_{PB}^{CP}(\tau^*) \cdot LV_{NPL}^{CP}(\tau^*) < C^{PB}(\tau^*),$$

and for $C_P^{LL}(\tau^*) > 0$ we get

$$\hat{F}_P^{LL}(\tau^*) = k_{LP}^{CP}(\tau^*) \cdot LV_{NPL}^{CP}(\tau^*) < C_P^{LL}(\tau^*).$$

The statement with respect to single Pfandbriefe follows from the definition of $\hat{F}_k^{PB}(t)$ for $k = 1, ..., n_{OL}$.

(c) If both the bank and the cover pool default due to overindebtedness at time $\hat{\tau} \in \mathcal{T}$, $\hat{\tau} = \tau = \tau^*$, it holds that

$$\begin{aligned} LV(\hat{\tau}) &\leq V^{CPS}(\hat{\tau}) + V^{OA}(\hat{\tau}) + N^{CPL}(\hat{\tau}) \\ &< B_{PB}^O(\hat{\tau}) + B_{OL}^O(\hat{\tau}) + N_B^{LL}(\hat{\tau}) \\ &\leq C^{PB}(\hat{\tau}) + C^{OL}(\hat{\tau}) + C_B^{LL}(\hat{\tau}) \end{aligned}$$

and

$$LV_{NPL}^{CP}(\hat{\tau}) \leq V_{NPL}^{CPS}(\hat{\tau}) + N^{CPL}(\hat{\tau}) < B_{PB}^{O}(\hat{\tau}) \leq C^{PB}(\hat{\tau})$$

as $h_{LI}^{OA}, h_{LI}^{CPS} \in [0,1]$ and $f_O(t,s) \leq 1 \; \forall s \geq t$. Furthermore, with

$$LV^{IE}(\hat{\tau}) = LV_{NPL}^{OA}(\hat{\tau}) + E^{PA}(\hat{\tau}),$$
$$C^{IE}(\hat{\tau}) = C^{OL}(\hat{\tau}) + \underbrace{C^{PB}(\hat{\tau}) - LV_{NPL}^{CP}(\hat{\tau})}_{G^{CP}(\hat{\tau})} + G_B^{LL}(\hat{\tau}),$$

and

$$E^{PA}(\hat{\tau}) - G_B^{LL}(\hat{\tau}) = LV_{PL}^{OA}(\hat{\tau}) + LV_{PL}^{CP}(\hat{\tau}) - C_B^{LL}(\hat{\tau}) \qquad (4.28)$$

it follows that

$$LV^{IE}(\hat{\tau}) - C^{IE}(\hat{\tau}) = LV(\hat{\tau}) - C^{PB}(\hat{\tau}) - C^{OL}(\hat{\tau}) - C_B^{LL}(\hat{\tau}) < 0.$$

Again, (4.28) holds because of $k_{LB}^{PA}(\hat{\tau}) \in \{0,1\}$. With $C^{PB}(\hat{\tau}) > 0$ and $C^{OL}(\hat{\tau}) > 0$, the liquidation payments to Pfandbrief holders and creditors of other liabilities are given by

$$\hat{F}^{PB}(\hat{\tau}) = LV_{NPL}^{CP}(\hat{\tau}) + k_{PB}^{G}(\hat{\tau}) \cdot LV^{IE}(\hat{\tau}) < C^{PB}(\hat{\tau}),$$
$$\hat{F}^{OL}(\hat{\tau}) = k_{OL}^{G}(\hat{\tau}) \cdot LV^{IE}(\hat{\tau}) < C^{OL}(\hat{\tau}),$$

and for $C_B^{LL}(\hat{\tau}) > 0$ we get

$$\hat{F}_B^{LL}(\hat{\tau}) = LV_{PL}^{OA}(\hat{\tau}) + LV_{PL}^{CP}(\hat{\tau}) + k_{LB}^{G}(\hat{\tau}) \cdot LV^{IE}(\hat{\tau}) < C_B^{LL}(\hat{\tau}).$$

The statement with respect to single Pfandbriefe and other liabilities follows from the definitions of $\hat{F}_k^{PB}(t)$ and $\hat{F}_l^{OL}(t)$ for $k = 1, ..., n_{PB}$ and $l = 1, ..., n_{OL}$. $\qquad\square$

In the special case where $t = T_{\max}$, there is a close link between overindebtedness and the ability to fulfil payment obligations.

Remark 4.34 (Overindebtedness in $T_{\max}$)
At time $t = T_{\max}$, the following statements are true:

(a) (OB) holds $\Leftrightarrow LV(T_{\max}) < F_B^L(T_{\max})$.

(b) (OP) holds $\Leftrightarrow LV_{NPL}^{CP}(T_{\max}) < F_P^L(T_{\max})$.

Proof. The first statement follows from $T_k^{PB}, T_l^{OL} \leq T_{\max}$ for $k = 1, ..., n_{PB}$ and $l = 1, ..., n_{OL}$ in conjunction with the zero coupon bond assumption and the fact that no liquidation haircuts apply in $T_{\max}$:

$$B_B^O(T_{\max}) = N^{PB}(T_{\max}) + N^{OL}(T_{\max}) + N_B^{LL}(T_{\max}) = F_B^L(T_{\max}),$$
$$LV(T_{\max}) = V^{CPS}(T_{\max}) + V^{OA}(T_{\max}) + N^{CPL}(T_{\max}) = V_B^O(T_{\max}).$$

Similarly, the second statement follows with

$$B_P^O(T_{\max}) = N^{PB}(T_{\max}) + N_P^{LL}(T_{\max}) = F_P^L(T_{\max}),$$
$$LV_{NPL}^{CP}(T_{\max}) = V_{NPL}^{CPS}(T_{\max}) + N^{CPL}(T_{\max}) = V_P^O(T_{\max}).$$

$\square$

4.6 Asset Liability Management

In the following, we specify the last component of the Pfandbrief model: the bank's and the cover pool administrator's asset liability management strategies in the context of funding, reinvestments, the maintenance of overcollateralization and liability payments (Sections 4.6.1–4.6.4). We do so by considering the required asset liability management features in Table 2.2. The updated balance sheet for the next simulation time step, which takes into account all these activities, is derived in Section 4.6.5.

4.6.1 Funding Strategies

Given our run-off assumption, which implies that no new business is made, funding is mainly needed to overcome temporary cash shortages. With respect to the bank's funding strategy, we therefore make the following assumption:

Assumption 25 (The bank's funding strategy)
At time $t \in \mathcal{T}$, $t \leq \tau$, the bank's funding strategy is as follows:

(a) The bank raises as much funding as necessary to fulfil its payment obligations and to maintain cover requirements but not more than that.

(b) As far as possible, the bank pledges other assets as collateral. Strategic cover pool assets are only pledged if there are not sufficient other assets available. Within each of these two asset classes, asset pledging is done pro rata.

(c) No funding is raised at time $t \in \{\tau, T_{\max}\}$.

The assumption that the bank pledges other assets with a higher priority than strategic cover pool assets is to be interpreted as the bank maintaining voluntary overcollateralization as long as possible. In a situation of funding pressure with no more other assets being available for pledging, the bank has no other choice but to reduce strategic cover pool assets in order to raise funding. Our assumptions regarding the cover pool administrator's funding strategy are similar but take into consideration that cover requirements are not actively maintained any more.

Assumption 26 (The cover pool administrator's funding strategy)
At time $t \in \mathcal{T}$, $\tau \leq t \leq \tau^*$, the cover pool administrator's funding strategy is as follows:

(a) The cover pool administrator raises as much funding as necessary to fulfil payment obligations but not more than that.

(b) No funding is raised at time $t \in \{\tau^*, T_{\max}\}$.

Given the above assumptions, the amount of funding raised and the fraction of pledged assets at a certain time t are fully specified.

Observation 4.35 (Funding activities)
At time $t \in \mathcal{T}$, the total amount of funding raised is

$$FR(t) := \begin{cases} G_B^C(t), & \text{if } S(t) = 1, \\ G_P^C(t), & \text{if } S(t) \in \{2, 4\}, \\ 0, & \text{otherwise}, \end{cases}$$

and the fractions of pledged assets are

$$k_{PL}^{OA}(t) := \begin{cases} \frac{\min(G_B^C(t); L_{OA}^{\max}(t))}{L_{OA}^{\max}(t)}, & \text{if } S(t) = 1,\ L_{OA}^{\max}(t) \neq 0, \\ 0, & \text{otherwise}, \end{cases}$$

and

$$k_{PL}^{CPS}(t) := \begin{cases} \dfrac{\min\left(\max(0; G_B^R(t)); L_{CPS}^{max}(t) \cdot z_B^{CPS}(t)\right)}{L_{CPS}^{max}(t)}, & \text{if } S(t) = 1, \ L_{CPS}^{max}(t) \neq 0, \\ 0, & \text{otherwise,} \end{cases}$$

with $G_B^R(t) := G_B^C(t) - L_{OA}^{max}(t)$.

Proof. The statement regarding $FR(t)$ follows straight from Assumptions 25 and 26 in conjunction with the definition of $S(t)$. As asset pledging is only relevant in the context of bank funding activities, cf. Assumptions 13 and 15, the fraction of pledged assets is non-zero only for $S(t) = 1$. Furthermore, according to Observation 4.22, the maximum amount of funding obtainable from pledging risky assets pro rata is given by $L_{OA}^{max}(t)$ and $L_{CPS}^{max} \cdot z_B^{CPS}(t)$ respectively. As other assets are pledged with a higher priority, their total pledged amount is $\min(G_B^C(t); L_{OA}^{max}(t))$, while the total amount of pledged strategic cover pool assets is $\min\left(\max(0; G_B^C(t) - L_{OA}^{max}(t)); L_{CPS}^{max} \cdot z_B^{CPS}(t)\right)$. The statements regarding $k_{PL}^{OA}(t)$ and $k_{PL}^{CPS}(t)$ then follow from the fact that within each of the two asset classes asset pledging is done pro rata. $\square$

Remark 4.36 (Funding activities in t_0)
At time t_0, it holds that $FR(t_0) = k_{PL}^{OA}(t_0) = k_{PL}^{CPS}(t_0) = 0$.

Proof. According to Assumption 17, the bank is not overindebted in t_0 and because of Assumptions 1, 12 and 14 it is also not illiquid:

$$G_B^C(t_0) = \max(0; F_B^L(t_0) + CR(t_0) - \tilde{F}_B^A(t_0)) = \max(0; -N^{CPL}(t_0)) = 0.$$

It follows that $S(t_0) = 1$ and $FR(t_0) = G_B^C(t_0) = 0$. The statement regarding $k_{PL}^{OA}(t_0)$ and $k_{PL}^{CPS}(t_0)$ is then straightforward. $\square$

4.6.2 Reinvestment Strategies

Given our run-off assumption, asset purchases only take place under certain circumstances, cf. Assumption 4. The reinvestment of excess cash inflows is one of these exceptions.

Definition 4.37 (Excess cash inflow of the bank)
The bank's excess cash inflow at time $t \in \mathcal{T}$ is given by

$$E_B^C(t) := \begin{cases} \max(0; \tilde{F}_B^A(t) - F_B^L(t) - CR(t)), & \text{if } t < T_{\max}, \ t \leq \tau, \\ 0, & \text{otherwise,} \end{cases}$$

with $\tilde{F}_B^A(t)$, $F_B^L(t)$ and $CR(t)$ as in Definition 4.21.

Excess cash inflows arise when the bank's realized time-t cash inflows exceed the amount of cash needed to fulfil payment obligations and to maintain cover requirements. As the simulation stops at time $T_{\max}$ and the balance sheet is liquidated, aspects in the context of reinvestments are not relevant any more at that time and we set $E_B^C(T_{\max}) = 0$.

With respect to the bank's reinvestment strategy, we make the following assumption:

Assumption 27 (Reinvestment strategy of the bank)
At time $t \in \mathcal{T}$, $t \leq \tau$, the bank's reinvestment strategy is as follows:

(a) The bank reinvests its excess cash inflow in other assets by increasing existing positions pro rata.

(b) No reinvestment activities take place at time $t \in \{\tau, T_{\max}\}$.

(c) At time t_0, excess cash inflows are kept as liquid cover pool assets.

Assuming that excess cash inflows are reinvested in other assets accounts for the fact that under the run-off assumption the incentive to hold or even increase voluntary overcollateralization might be rather low as reputational aspects, target ratings and program support lose relevance, cf. Langer and Schadow [109]. Our assumption is therefore conservative and reflects the voluntary nature of overcollateralization in excess of legal requirements. The pro rata increase of existing other asset positions ensures that the reinvestment strategy does not have a biasing impact on asset credit risk and asset liability mismatches.

In the cover pool's case, excess cash inflows arise when time-t cover pool cash inflows exceed the amount of cash needed to fulfil payment obligations; cover requirements are not taken into consideration any more.

Definition 4.38 (Excess cash inflow of the cover pool)
The cover pool's excess cash inflow at time $t \in \mathcal{T}$ is given by

$$
E_P^C(t) := \begin{cases} \max(0; \tilde{F}_P^A(t) - F_P^L(t)), & \text{if } t < T_{\max}, \ \tau \leq t \leq \tau^*, \\ 0, & \text{otherwise,} \end{cases}
$$

with $F_P^A(t) := \tilde{F}_{NPL}^{CPS}(t) + N^{CPL}(t)$ and $F_P^L(t) = F^{PB}(t) + N_P^{LL}(t)$, where $\tilde{F}_{NPL}^{CPS}(t)$ as in Definition 4.23.

The reinvestment strategy of the cover pool administrator is assumed to be different from the bank's one. For an operating bank, the accumulation of large amounts of cash is not very efficient, but in the special case of a standalone (potentially illiquid) cover pool[17] it can be argued that the cover pool administrator acts prudently and tries to avoid future cash shortages by all means. As cover requirements, especially the 180-day liquidity rule, are not maintained any more, the accumulation of cash positions is therefore a reasonable assumption. Alternatively, one could specify more sophisticated reinvestment strategies which aim at minimizing cash holdings while still ensuring a reasonable cash buffer at all times. The definition of such strategies is, however, beyond the scope of this work.

Assumption 28 (Reinvestment strategy of the cover pool)
At time $t \in \mathcal{T}$, $\tau \leq t \leq \tau^*$, the cover pool administrator's reinvestment strategy is as follows:

(a) He reinvests the cover pool's excess cash inflow in liquid cover pool assets by increasing the corresponding position accordingly.

(b) No reinvestment takes place at time $t \in \{\tau^*, T_{\max}\}$.

The reinvestment activities of the bank and the cover pool administrator are summarized by Observation 4.39 below.

Observation 4.39 (Reinvestment activities)
At time $t \in \mathcal{T}$, the fraction by which other assets are increased due to

---

[17] As already mentioned before, we concentrate on Mortgage Pfandbrief modelling.

reinvestment activities is

$$k_{inc}^{OA}(t) := \begin{cases} \frac{E_B^C(t)}{\tilde{V}^{OA}(t)}, & \text{if } S(t) = 1,\ t > t_0, \\ 0, & \text{otherwise,} \end{cases}$$

and the amount by which liquid cover pool assets are increased is

$$m_{inc}^{CPL}(t) := \begin{cases} E_B^C(t), & \text{if } S(t) = 1,\ t = t_0, \\ E_P^C(t), & \text{if } S(t) \in \{2,4\}, \\ 0, & \text{otherwise.} \end{cases}$$

Strategic cover pool assets are not increased due to reinvestment activities.

Proof. The statement regarding $m_{inc}^{CPL}(t)$ follows straight from Assumptions 27 and 28. Other assets are only increased if $S(t) = 1$ and $t > t_0$, and in this case they are increased by an amount of $E_B^C(t)$. As the reinvestment is done pro rata, the corresponding fraction by which other assets need to be increased is given by $k_{inc}^{OA}(t)$. It remains to show that $k_{inc}^{OA}(t)$ is well-defined, meaning that $\tilde{V}^{OA}(t) \neq 0$ for $t_0 < t < \min(\tau, T_{\max})$. Due to Assumptions 1 and 2, it holds that $N_{n_{OA}}^{OA}(t_0) > 0$ and $T_{n_{OA}}^{OA} \geq T_{\max}$. In our model, assets cannot default prior to maturity and the nominal of the n_{OA}-th other asset cannot be reduced to zero through asset sales as long as $t < \min(\tau, T_{\max})$, cf. Assumption 4. We therefore get $\tilde{N}_{n_{OA}}^{OA}(t) > 0$. In addition to that, our asset model ensures that $\tilde{V}_{n_{OA},U}^{OA}(t) > 0$ for $t < T_{\max} \leq T_{n_{OA}}^{OA}$, cf. Remark 4.15. All in all, $\tilde{V}^{OA}(t) > 0$ for $t < \min(\tau, T_{\max})$. □

As the proof of Observation 4.39 reveals, a relaxation of Assumption 1 with asset defaults being possible prior to maturity would result in a need for more sophisticated reinvestment strategies as for $t_0 < t < \min(\tau, T_{\max})$ it could not be guaranteed any more that there are always non-defaulted other assets available for pro rata reinvestments.

Remark 4.40 (Reinvestment activities in t_0)
At time t_0, it holds that $k_{inc}^{OA}(t_0) = 0$ and $m_{inc}^{CPL}(t_0) = N^{CPL}(t_0)$.

Proof. From the proof of Remark 4.36 we know that $S(t_0) = 1$. From Observation 4.39 we have $k_{inc}^{OA}(t) = 0$, and from Assumptions 1, 12, 14 we get $m_{inc}^{CPL}(t_0) = E_B^C(t_0) = \max(0; \tilde{F}_B^A(t_0) - CR(t_0) - F_B^L(t_0)) = N^{CPL}(t_0)$. □

4.6.3 Maintenance of Overcollateralization

Apart from the reinvestment of excess cash inflows, there is only one other situation under which asset purchases take place given our run-off assumption. The legal cover requirements (C1)–(C3) have to be maintained by the bank at all times and immediate action has to be taken whenever the matching cover calculations reveal gaps. In this context, we make the following assumption:

Assumption 29 (Reestablishment of legal cover requirements)
At time $t \in \mathcal{T}$, $t \leq \tau$, the bank's strategy to reestablish the legal cover requirements is as follows:

(a) The bank first makes sure that the nominal cover requirement (C1) is fulfilled, if necessary by increasing strategic cover pool assets pro rata by an amount of

$$G_{CR}^{CPS}(t) := k_{inc}^{CPS}(t) \cdot \tilde{V}^{CPS}(t)$$

through asset purchases, where $\tilde{k}_{inc}^{CPS}(t) := \frac{\max(0; \tilde{N}^{PB}(t) - \tilde{N}^{CPS}(t))}{\tilde{N}^{CPS}(t)}$.

(b) Having reestablished the nominal cover requirement, the bank takes care of the cover requirements (C2) and (C3) by adding liquid cover pool assets of an amount

$$G_{CR}^{CPL}(t) := \max(G_{CR}^{C2}(t); G_{CR}^{C3}(t)),$$

with

$$G_{CR}^{C2}(t) := \max(G_{CR}^{C2,b}(t); G_{CR}^{C2,d}(t); G_{CR}^{C2,u}(t)),$$
$$G_{CR}^{C3}(t) := \max_{s \in \mathcal{T}_{CPS} \cup \mathcal{T}_{PB}, \ t < s \leq t+0.5} F_{cum}^{net}(t,s),$$
$$G_{CR}^{C2,y}(t) := \max(0; 1.02 \cdot \tilde{V}_{*y}^{PB}(t) - (1 + k_{inc}^{CPS}(t)) \cdot \tilde{V}_{*y}^{CPS}(t)),$$
$$F_{cum}^{net}(t,s) := \max(0; F_{cum}^{PB}(t,s) - (1 + k_{inc}^{CPS}(t)) \cdot F_{cum}^{CPS}(t,s)),$$

for $y \in \{b, u, d\}$.

(c) No cover requirements are reestablished in $t \in \{\tau, T_{\max}\}$.

According to the above assumption, additional strategic cover pool assets are added when the nominal cover requirement is breached. This is in line with the PfandBG, which requires additional eligible assets to be posted in this case. Liquid cover pool assets could also be used for this purpose (as further cover assets) but only within certain restrictions. Allowing for additional liquid cover pool assets would require a distinction of different asset liability management strategies based on the current composition of the cover pool to ensure that the restrictions regarding the amount of further cover pool assets are not violated when the nominal cover is reestablished. The posting of additional liquid cover pool assets to cure a breach of the nominal cover requirement is, however, not likely to be the bank's preferred choice as holding more cash than needed and legally required is not profitable. The assumption that new strategic cover pool assets are bought instead of transferring existing other assets to the cover pool avoids further model complexity arising from the need to distinguish additional asset subcategories (cover pool eligible other assets vs. non-eligible other assets) and related asset liability management strategies. A pro-rata increase of strategic cover pool assets ensures that the cover pool's composition and default risk characteristics remain unchanged.[18] The fraction $\tilde{k}_{inc}^{CPS}(t) = 0$ is well-defined for $t < T_{\max}$ and $t \leq \tau$. This is due to the facts that $N_{n_{CPS}}^{CPS}(t_0) > 0$, defaults are not possible prior to maturity, $T_{n_{CPS}}^{CPS} \geq T_{\max}$ and asset sales cannot occur prior to time $t < T_{\max}$ as long as $t \leq \tau$ (cf. Assumptions 1, 2, 4), which implies $\tilde{N}^{CPS}(t) > 0$. Furthermore, Assumption 29 states that additional liquid cover pool assets are posted if one of the cover requirements in (C2) and (C3) is violated. This is in line with the PfandBG which stipulates that highly liquid assets need to be added in this case. Our assumption that the reestablishment of cover requirements is done in two steps ensures an unambiguous specification of formulas according to which the corresponding positions are determined. At time τ (when the bank defaults) and in $T_{\max}$ (when the simulation stops and the balance sheet is liquidated), cover requirements are not maintained any more.

Note that Assumption 29 does not make any statement with respect to BaFin's competence to increase the required overcollateralization for individual Pfandbrief programmes if deemed necessary (cover add-on). Due to the fact that such a decision would be completely at the discretion of BaFin, with timing and size being difficult to anticipate, this additional buffer is currently ignored in our model. It could, however, be incorporated by extending asset liability management strategies accordingly, which in turn requires additional (strong) assumptions to be made.

---

[18]The assumption that reinvestment activities concentrate on assets with similar credit characteristics is also made by DBRS [37].

Observation 4.41 (Reestablishment of legal cover requirements)
At time $t \in \mathcal{T}$, $t \le \tau$, the amount of cash needed to reestablish the cover requirements is given by

$$CR(t) := \begin{cases} G_{CR}^{CPS}(t) + G_{CR}^{CPL}(t), & \text{if } t < T_{\max}, \\ 0, & \text{otherwise.} \end{cases}$$

Proof. The statement follows from the two-step strategy in Assumption 29. $\square$

In practice, Pfandbrief banks tend to maintain voluntary overcollateralization, but as already discussed before the incentive to hold such voluntary overcollateralization can be expected to be rather low under our run-off assumption.

Assumption 30 (Maintenance of overcollateralization by the bank)
At time $t \in \mathcal{T}$, $t \le \tau$, the bank's strategy to maintain overcollateralization is as follows:

(a) *The bank never adds more assets to the cover pool than needed to reestablish the legal cover requirements (C1)–(C3).*

(b) *Strategic cover pool assets in excess of the nominal cover requirement (C1) remain in the cover pool as long as not needed for funding purposes.*

(c) *Liquid cover pool assets are withdrawn immediately once not needed any more to cure a breach of the cover requirements (C2) and (C3).*

(d) *No overcollateralization is maintained at time $t \in \{\tau, T_{\max}\}$.*

According to the above assumption, the bank ensures that the legal cover requirements are fulfilled at all times, but voluntary overcollateralization is not actively maintained. There may be voluntary overcollateralization in terms of (C1), but once the issuer needs additional funding these strategic cover pool assets can be pledged as collateral, implying that voluntary overcollateralization is (temporary) reduced up to the legal minimum. With respect to (C2) and (C3), no voluntary overcollateralization is held at all, i.e. liquid cover pool assets are only used for liquidity steering if one of these two cover requirements is breached. No overcollateralization is maintained at time τ (when the bank defaults) and in $T_{\max}$ (when the simulation stops and the balance sheet is liquidated).

The cover pool administrator has no strategy regarding the maintenance of overcollateralization as he only administers the cover pool.[19]

Assumption 31 (Maintenance of overcollateralization by the cover pool administrator)
At time $t \in \mathcal{T}$, $\tau \leq t \leq \tau^*$, the cover pool administrator does not actively maintain overcollateralization.

The activities in the context of maintenance of overcollateralization can be summarized as follows.

Observation 4.42 (Maintenance of overcollateralization)
The present value of assets which are added to the cover pool at time $t \in \mathcal{T}$ to maintain overcollateralization is

$$AA(t) := \begin{cases} CR(t), & \text{if } S(t) = 1, \\ 0, & \text{otherwise,} \end{cases}$$

and the amount by which strategic cover pool assets are increased through asset purchases is

$$k_{inc}^{CPS}(t) := \begin{cases} \tilde{k}_{inc}^{CPS}(t), & \text{if } S(t) = 1, \\ 0, & \text{otherwise.} \end{cases}$$

Proof. The statement follows from Assumptions 29, 30, 31 and Observation 4.41. $\qquad\square$

---

[19]Even though the PfandBG stipulates that the cover pool administrator has to check the cover requirements on a regular basis, cf. PfandBG § 31 par. 5, he does not have access to additional assets which he could post to the cover pool once one of the cover requirements is breached.

4.6.4 Liability Payments

Liability payments depend on the occurrence of the events of bank and cover pool default and the simulation time t. In line with our previous discussions which are summarized by Table 4.2, we make the following assumption:

Assumption 32 (Liability payments)
At time $t \in \mathcal{T}$, liability payments take place according to the following principles:

(a) As long as no event of bank default occurs and $t < T_{\max}$, the creditors of other liabilities and the bank's liquidity line receive scheduled payments.

(b) Pfandbrief holders and the cover pool's liquidity line receive scheduled payments as long as no event of cover pool default occurs and $t < T_{\max}$.

(c) In $T_{\max}$, liquidation payments are made according to Observation 4.30.

(d) Apart from the cases as specified by (a)–(c), no liability payments take place.

Given the above assumption, the determination of realized liability payments, which do not necessarily equal contractually scheduled ones, is straightforward.

Observation 4.43 (Realized liability payments)
At time $t \in \mathcal{T}$, realized liability payments are given by

$$\tilde{F}_k^{PB}(t) := \begin{cases} N_k^{PB}(t_0), & \text{if } S(t) \in \{1, 2, 4\} \text{ and } t = T_k^{PB}, \\ \hat{F}_k^{PB}(t), & \text{if } S(t) \in \{3, 5, 7, 8\}, \\ 0, & \text{otherwise,} \end{cases}$$

$$\tilde{F}_l^{OL}(t) := \begin{cases} N_l^{OL}(t_0), & \text{if } S(t) = 1 \text{ and } t = T_l^{OL}, \\ \hat{F}_l^{OL}(t), & \text{if } S(t) \in \{2, 3, 5, 7, 8\}, \\ 0, & \text{otherwise,} \end{cases}$$

$$\tilde{F}_B^{LL}(t) := \begin{cases} \max\left(0; N_B^{LL}(t) - G_B^C(t)\right), & \text{if } S(t) = 1, \\ \hat{F}_B^{LL}(t), & \text{if } S(t) \in \{2, 3, 5, 7, 8\}, \\ 0, & \text{otherwise,} \end{cases}$$

$$\tilde{F}_P^{LL}(t) := \begin{cases} \max\left(0; N_P^{LL}(t) - G_P^C(t)\right), & \text{if } S(t) = 4, \\ \hat{F}_P^{LL}(t), & \text{if } S(t) \in \{5, 8\}, \\ 0, & \text{otherwise}, \end{cases}$$

$$\tilde{F}^{EQ}(t) := \begin{cases} \hat{F}^{EQ}(t), & \text{if } S(t) \in \{2, 3, 5, 7, 8\} \\ 0, & \text{otherwise}, \end{cases}$$

for $k = 1, ..., n_{PB}$ and $l = 1, ..., n_{OL}$. The total payments to Pfandbrief holders and creditors of other liabilities are given by

$$\tilde{F}^{PB}(t) := \sum_{k=1}^{n_{PB}} \tilde{F}_k^{PB}(t) \quad \text{and} \quad \tilde{F}^{OL}(t) := \sum_{l=1}^{n_{OL}} \tilde{F}_l^{OL}(t).$$

Proof. As Pfandbriefe and other liabilities are modelled as zero coupon bonds, scheduled payments to the corresponding creditors are given by $N_k^{PB}(t_0)$ in T_k^{PB} and by $N_l^{OL}(t_0)$ in T_l^{OL}, with $k = 1, ..., n_{PB}$ and $l = 1, ..., n_{OL}$. The scheduled payments to the liquidity lines correspond to the net cash outflows $\max\left(0; N_x^{LL}(t) - G_x^C(t)\right)$, $x \in \{B, P\}$ and take into account the payment $N_x^{LL}(t)$ which is due at time t as well as potential new funding $G_x^C(t)$ raised at that time. Equity holders, on the other hand, do not receive any scheduled dividend payments, cf. Assumption 1. The scenarios $S(t)$ under which scheduled payments take place follow straight from Assumption 32. Liquidation payments only occur for $S(t) \in \{2, 3, 5, 7, 8\}$ and the affected liabilities are those for which the liquidation claims according to Assumption 23 are bigger than zero. The liquidation payments are then given by $\hat{F}_k^{PB}(t)$, $\hat{F}_l^{OL}(t)$, $\hat{F}_B^{LL}(t)$ and $\hat{F}_P^{LL}(t)$ as in Observation 4.30. □

Remark 4.44 (Realized liability payments in t_0)
There are no realized liability payments in t_0:

$$\tilde{F}_k^{PB}(t_0) = \tilde{F}_l^{OL}(t_0) = \tilde{F}_B^{LL}(t_0) = \tilde{F}_P^{LL}(t_0) = \tilde{F}^{EQ}(t_0) = 0,$$

for $k = 1, ..., n_{CPS}$ and $l = 1, ..., n_{OA}$.

Proof. From the proof of Remark 4.36 we know that $S(t_0) = 1$ and according to Assumptions 1, 14 and 16 there are no scheduled liability payments in t_0. □

4.6.5 Balance Sheet Update

Due to market movements, dynamic asset liability management and potential default events as specified in the previous sections, the balance sheet changes stochastically over time. The associated balance sheet update from time t to time $t + \Delta$ is reflected in the following formulas:

Observation 4.45 (Updated balance sheet)
Given the balance sheet at time $t \in \mathcal{T}$, $t < T_{\max}$, the updated asset nominals at time $t + \Delta$ after all asset liability management activities are

$$N_i^{CPS}(t + \Delta) := \begin{cases} N_i^{CPS}(t) \cdot \left(1 + k_{inc}^{CPS}(t)\right), & \text{if } S(t) = 1 \text{ and } t < T_i^{CPS}, \\ N_{i,NPL}^{CPS}(t), & \text{if } S(t) = 2 \text{ and } t < T_i^{CPS}, \\ N_i^{CPS}(t), & \text{if } S(t) = 4 \text{ and } t < T_i^{CPS}, \\ 0, & \text{otherwise,} \end{cases}$$

$$N^{CPL}(t + \Delta) := \begin{cases} \frac{G_{CR}^{CPL}(t) + m_{inc}^{CPL}(t)}{P(t,t+\Delta)}, & \text{if } S(t) = 1, \\ \frac{m_{inc}^{CPL}(t)}{P(t,t+\Delta)}, & \text{if } S(t) \in \{2, 4\}, \\ 0, & \text{otherwise,} \end{cases}$$

$$N_j^{OA}(t + \Delta) := \begin{cases} N_j^{OA}(t) \cdot \left(1 + k_{inc}^{OA}(t)\right), & \text{if } S(t) = 1 \text{ and } t < T_j^{OA}, \\ 0, & \text{otherwise,} \end{cases}$$

for $i = 1, ..., n_{CPS}$ and $j = 1, ..., n_{OA}$, with $k_{inc}^{CPS}(t)$ and $m_{inc}^{CPL}(t)$ as in Observation 4.39. For pledged assets it holds that

$$N_{i,PL}^{CPS}(t + \Delta) := \begin{cases} k_{PL}^{CPS}(t) \cdot N_i^{CPS}(t), & \text{if } S(t) = 1 \text{ and } t < T_i^{PB}, \\ 0, & \text{otherwise,} \end{cases}$$

$$N_{j,PL}^{OA}(t + \Delta) := \begin{cases} k_{PL}^{OA}(t) \cdot N_j^{OA}(t), & \text{if } S(t) = 1 \text{ and } t < T_j^{OL}, \\ 0, & \text{otherwise,} \end{cases}$$

with $k_{PL}^{CPS}(t)$ and $k_{PL}^{OA}(t)$ as in Observation 4.35.

Furthermore, the updated liability nominals are

$$N_k^{PB}(t + \Delta) := \begin{cases} N_k^{PB}(t), & \text{if } S(t) \in \{1, 2, 4\} \text{ and } t < T_k^{PB}, \\ 0, & \text{otherwise,} \end{cases}$$

$$N_l^{OL}(t + \Delta) := \begin{cases} N_l^{OL}(t) & \text{if } S(t) = 1 \text{ and } t < T_l^{OL}, \\ 0, & \text{otherwise}, \end{cases}$$

$$N_B^{LL}(t + \Delta) := \begin{cases} G_B^C(t) \cdot P(t, t + \Delta)^{-1}, & \text{if } S(t) = 1, \\ 0, & \text{otherwise}, \end{cases}$$

$$N_P^{LL}(t + \Delta) := \begin{cases} G_P^C(t) \cdot P(t, t + \Delta)^{-1}, & \text{if } S(t) \in \{2, 4\}, \\ 0, & \text{otherwise}, \end{cases}$$

for $k = 1, ..., n_{PB}$ and $l = 1, ..., n_{OL}$

Proof. The statement follows from Assumption 1 and the asset liability management strategies as specified in Sections 4.6.1–4.6.4. □

4.7 Default Parameters

Having specified all components of our Pfandbrief model in the previous sections of this chapter, we now derive the liabilities' default parameters. According to Assumption 19, a bank default at time τ triggers the acceleration of the bank's outstanding other liabilities which become due and payable immediately. The creditor of the l-th other liability, $l = 1, ..., n_{OL}$, is affected by this event if $\tau \leq T_l^{OL}$. Similarly, the fate of the bank's outstanding Pfandbriefe depends on the occurrence of an event of cover pool default, cf. Assumption 21, and the k-th Pfandbrief holder, $k = 1, ..., n_{PB}$, is affected by this event if $\tau^* \leq T_k^{PB}$. The derivation of default probabilities is straightforward:

Observation 4.46 (Default probabilities)
At time $t \in \mathcal{T}$, the probability of a bank default is $\pi_t^B := \mathbb{P}(\tau = t)$ and the probability of a cover pool default is $\pi_t^P := \mathbb{P}(\tau^* = t)$. The corresponding cumulative default probabilities are $\pi_{t,cum}^B := \mathbb{P}(\tau \leq t)$ and $\pi_{t,cum}^P := \mathbb{P}(\tau^* \leq t)$.

Proof. The above statements follow from Assumptions 18 and 20. □

Under the recovery of treasury assumption, which stipulates that a defaulted bond is replaced by a risk-free bond with the same maturity but with a reduced nominal, the associated losses given default can be derived as follows.

Observation 4.47 (Losses given default)
In case the cover pool defaults at time $t \in \mathcal{T}$, the loss given default of outstanding Pfandbriefe is

$$L_t^{PB} := 1 - \mathbb{E}^{\mathbb{P}}\left[\left.\frac{\hat{F}^{PB}(\tau^*)}{C^{PB}(\tau^*)}\right| \tau^* = t\right].$$

Similarly, if the bank defaults at time $t \in \mathcal{T}$, the loss given default of outstanding other liabilities is

$$L_t^{OL} := 1 - \mathbb{E}^{\mathbb{P}}\left[\left.\frac{\hat{F}^{OL}(\tau) + PV_\tau\left[\bar{F}^{OL}(\hat{\tau})\right]}{C^{OL}(\tau)}\right| \tau = t\right].$$

Here, $\hat{\tau} = \min\left(\tau^*, T_{\max}\right)$ and

$$\bar{F}^{OL}(\hat{\tau}) := \begin{cases} \hat{F}^{OL}(\hat{\tau}), & \text{if } \hat{\tau} > \tau, \\ 0, & \text{otherwise}, \end{cases}$$

with

$$PV_\tau\left[Z(\hat{\tau})\right] := \mathbb{E}^{\mathbb{Q}}\left[\left. e^{-\int_\tau^{\hat{\tau}} r(s)ds} \cdot Z(\hat{\tau})\right| \mathcal{F}_\tau\right], \tag{4.29}$$

for $Z \in \mathcal{L}^1(\Omega, \mathcal{F}, \mathbb{P})$.

Proof. According to Assumption 23, Pfandbrief holders can raise a claim of $C^{PB}(\tau^*)$ at the time $\tau^* \in \mathcal{T}$ of cover pool default and the received liquidation payment is given by $\hat{F}^{PB}(\tau^*)$, see Observation 4.30. Apart from the cash flows at time τ^*, no recovery payments take place. To derive the associated loss given default, the time-τ^* present value of the received liquidation payment, which is equal to the cash flow itself, has to be set in relation to the original claim. Conditional on $\tau^* = t$, the stochastic claim $C^{PB}(\tau^*)$ is bigger than zero, i.e. L_t^P is well defined.

In the case of the bank's other liabilities, the situation is a bit more complicated. At the time $\tau \in \mathcal{T}$ of bank default, the corresponding creditors can raise a claim of $C^{OL}(\tau)$ and, if $\tau < \hat{\tau}$ with $\hat{\tau} := \min(\tau^*, T_{\max})$, they can also raise a claim of $C^{OL}(\hat{\tau})$ at time $\hat{\tau}$, cf. Assumption 23. The received liquidation payments are given by $\hat{F}^{OL}(\tau)$ and $\bar{F}^{OL}(\hat{\tau})$, see Observation 4.30. To derive the associated loss given default, the time-τ present values of the received liquidation payments have to be set in relation to the original claim. In the case of $\hat{F}^{OL}(\tau)$, the time-τ present value of the cash flow is equal to the cash flow itself, but for the cash flow $\bar{F}^{OL}(\hat{\tau})$ the present value $PV_\tau\left[\bar{F}^{OL}(\hat{\tau})\right]$ has to be taken. Again, L_t^B is well defined as, conditional on $\tau = t$, the stochastic claim $C^{OL}(\tau)$ is bigger than zero.   □

The loss given default of Pfandbriefe as defined in Observation 4.47 correctly reflects the priority of payments from the Pfandbrief holder's perspective. In the case of other liabilities, the loss given default should be interpreted with care as there is no distinction between different levels of seniority (e.g. senior unsecured debt, subordinated debt) as typically observed in practice. We therefore consider the model's primary output to be Pfandbrief default statistics. The default statistics of other liabilities are only used for comparison.

The calculation of L_t^{OL} requires the evaluation of the risk neutral expectation in Equation (4.29). As there is no analytical expression for $PV_\tau\left[\bar{F}^{OL}(\hat{\tau})\right]$, additional risk neutral simulations are needed to price this cash flow in case of a bank default at time $\tau < \hat{\tau}$. To avoid additional model complexity, we make a simplifying assumption:

Assumption 33 (Approximation of loss given default (part I))
The Radon-Nikodym derivative $\frac{d\mathbb{Q}_{t_i}}{d\mathbb{P}}$ of the forward measure $\mathbb{Q}_{t_i}$, $i = 1, ..., S$, with respect to the real-world measure $\mathbb{P}$ is independent of $\mathbb{I}_{\{\hat{\tau}=t_i\}} \cdot \bar{F}_l^{OL}(t_i)$ and $\mathcal{F}_\tau$ for $l = 1, ..., n_{OL}$.

It can be shown, cf. Proposition 4.48 below, that under the above assumption it is sufficient to simulate additional real-world scenarios in case of a bank default at time $\tau < \hat{\tau}$, i.e. no risk neutral simulations are needed.

Proposition 4.48 (Calculation of losses given default)
Under Assumption 33, the time-τ present value $PV_\tau\left[\bar{F}_l^{OL}(\hat{\tau})\right]$ is given by

$$PV_\tau\left[\bar{F}^{OL}(\hat{\tau})\right] = P(\tau, \hat{\tau}) \cdot \mathbb{E}^\mathbb{P}\left[\bar{F}^{OL}(\hat{\tau})\big|\mathcal{F}_\tau\right], \quad l = 1, ..., n_{OL}.$$

Proof. The time-τ present value $PV_\tau\left[\bar{F}_l^{OL}(\hat{\tau})\right]$ can be rewritten as

$$PV_\tau\left[\bar{F}_l^{OL}(\hat{\tau})\right] = \sum_{i=1}^{S}\mathbb{E}^\mathbb{Q}\left[\mathbb{I}_{\{\hat{\tau}=t_i\}} \cdot e^{-\int_\tau^{t_i} r(s)ds} \cdot \bar{F}_l^{OL}(t_i)\bigg|\mathcal{F}_\tau\right],$$

and switching to the t_i-forward measures $\mathbb{Q}_{t_i}$ yields

$$PV_\tau \left[\bar{F}_l^{OL}(\hat{\tau}) \right] = \sum_{i=1}^{S} P(\tau, t_i) \cdot \mathbb{E}^{\mathbb{Q}_{t_i}} \left[\mathbb{I}_{\{\hat{\tau}=t_i\}} \cdot \bar{F}_l^{OL}(t_i) \Big| \mathcal{F}_\tau \right]$$

$$= \sum_{i=1}^{S} P(\tau, t_i) \cdot \mathbb{E}^{\mathbb{P}} \left[\frac{d\mathbb{Q}_{t_i}}{d\mathbb{P}} \cdot \mathbb{I}_{\{\hat{\tau}=t_i\}} \cdot \bar{F}_l^{OL}(t_i) \Big| \mathcal{F}_\tau \right].$$

Under the assumption that $\frac{d\mathbb{Q}_{t_i}}{d\mathbb{P}}$ is independent of $\mathcal{F}_\tau$ and $\mathbb{I}_{\{\hat{\tau}=t_i\}} \cdot \bar{F}_l^{OL}(t_i)$, it follows from the product rule for conditional expectation that

$$PV_\tau \left[\bar{F}_l^{OL}(\hat{\tau}) \right] = \sum_{i=1}^{S} P(\tau, t_i) \cdot \mathbb{E}^{\mathbb{P}} \left[\frac{d\mathbb{Q}_{t_i}}{d\mathbb{P}} \right] \cdot \mathbb{E}^{\mathbb{P}} \left[\mathbb{I}_{\{\hat{\tau}=t_i\}} \cdot \bar{F}_l^{OL}(t_i) \Big| \mathcal{F}_\tau \right],$$

and with $\mathbb{E}^{\mathbb{P}}[\frac{d\mathbb{Q}_{t_i}}{d\mathbb{P}}] = 1$ we get

$$PV_\tau \left[\bar{F}_l^{OL}(\hat{\tau}) \right] = P(\tau, \hat{\tau}) \cdot \mathbb{E}^{\mathbb{P}} \left[\bar{F}_l^{OL}(\hat{\tau}) \Big| \mathcal{F}_\tau \right]$$

for $l = 1, ..., n_{OL}$. It follows that

$$PV_\tau \left[\bar{F}^{OL}(\hat{\tau}) \right] = PV_\tau \left[\sum_{l=1}^{n_{OL}} \bar{F}_l^{OL}(\hat{\tau}) \right] = \sum_{l=1}^{n_{OL}} PV_\tau \left[\bar{F}_l^{OL}(\hat{\tau}) \right]$$

$$= \sum_{l=1}^{n_{OL}} P(\tau, \hat{\tau}) \cdot \mathbb{E}^{\mathbb{P}} \left[\bar{F}_l^{OL}(\hat{\tau}) \Big| \mathcal{F}_\tau \right] = P(\tau, \hat{\tau}) \cdot \mathbb{E}^{\mathbb{P}} \left[\bar{F}^{OL}(\hat{\tau}) \Big| \mathcal{F}_\tau \right].$$

$$\square$$

As our focus is on Pfandbrief modelling, i.e. on the determination of Pfandbrief default parameters (and not on those of other liabilities), it is acceptable to assume that the change of measure from $\mathbb{Q}_{t_i}$ to $\mathbb{P}$ is independent of $\mathcal{F}_\tau$, $\mathbb{I}_{\{\hat{\tau}=t_i\}} \cdot \bar{F}_l^{OL}(t_i)$. To avoid additional simulations, we use a one-path real-world simulation to approximate $\mathbb{E}^{\mathbb{P}} \left[\bar{F}^{OL}(\hat{\tau}) \big| \mathcal{F}_\tau \right] = \sum_{l=1}^{n} \mathbb{E}^{\mathbb{P}} \left[\bar{F}_l^{OL}(\hat{\tau}) \big| \mathcal{F}_\tau \right]$. This is expected to only have a minor impact on the loss given default of the other liabilities as the second liquidation cash flow should be comparably small.[20] More specifically, we make the following assumption:

---

[20]Indeed, in Section 6.2 we will find that simulation results obtained from an exemplary model calibration indicate that the value of the subordinated claim of other liabilities on the cover pool is rather small.

Assumption 34 (Approximation of loss given default (part II))
The real-world expectations $\mathbb{E}^{\mathbb{P}}\left[\bar{F}_l^{OL}(\hat{\tau})\middle|\mathcal{F}_\tau\right]$ *can be approximated by a one-path simulation under the real-world measure:*

$$\mathbb{E}^{\mathbb{P}}\left[\bar{F}_l^{OL}(\hat{\tau})\middle|\mathcal{F}_\tau\right] \approx \bar{F}_l^{OL}(\hat{\tau}), \quad l = 1, ..., n_{OL}.$$

Having specified the default probabilities and the losses given default, expected losses can be easily derived.

Definition 4.49 (Expected losses)
In case of a cover pool default at time $t \in \mathcal{T}$, the expected loss of outstanding Pfandbriefe is $EL_t^{PB} := \pi_t^P \cdot L_t^{PB}$. Similarly, in case of a bank default at time $t \in \mathcal{T}$, the expected loss of outstanding other liabilities is $EL_t^{OL} := \pi_t^B \cdot L_t^{OL}$. The corresponding cumulative expected losses are given by $EL_{t,cum}^{PB} := \sum_{s \in \mathcal{M}_t} \pi_s^P \cdot L_s^{PB}$ and $EL_{t,cum}^{OL} := \sum_{s \in \mathcal{M}_t} \pi_s^B \cdot L_s^{OL}$ with $\mathcal{M}_t := \{s \in \mathcal{T}, s \leq t\}$.

With the help of the above default parameters, we define average losses given default for the bank's Pfandbriefe and other liabilities.

Definition 4.50 (Average losses given default)
The average loss given default of the bank's Pfandbriefe is defined by

$$L_{avg}^{PB} := \frac{EL_{S,cum}^{PB}}{\pi_{S,cum}^P},$$

and the average loss given default of the bank's other liabilities is defined by

$$L_{avg}^{OL} := \frac{EL_{S,cum}^{OL}}{\pi_{S,cum}^B},$$

with $S = T_{\max}$ being the maturity of the longest Pfandbrief and the longest other liability.

All in all, we have now fully specified our Pfandbrief model which fulfils all modelling requirements in Table 2.2. To obtain meaningful default parameter estimates, the model needs to be adequately calibrated.

5 Model Calibration and Scenario Generation

This chapter is focused on an exemplary model calibration for Mortgage Pfandbriefe. In Section 5.1 we describe how the bank's balance sheet profile can be derived based on publicly available data and on expert opinion. Section 5.2 then addresses the calibration of the market environment, i.e. the determination of the parameters of the risk-free interest rate dynamics and the state variable processes of the bank's risky assets. In Section 5.3 the generation of the stochastic scenarios of the market environment is specified.

5.1 An Exemplary Balance Sheet Profile

The determination of the balance sheet parameters in Assumption 1 is not straightforward due to the lack of publicly available information. While data on nominals and maturities is available on an aggregated level in §28 Pfandbrief statistics and in the bank's annual reports, no issuer-specific data on asset default parameters (PDs, LGDs and correlations) and asset present values is typically published. For the subsequent analyses we derive an exemplary balance sheet by using published data from a large active issuer of Mortgage Pfandbriefe to approximate typical asset liability mismatches and choose average representative values for asset default parameters. To obtain realistic bank-specific results, more granular information on the balance sheet composition (as for example available to the issuing bank, rating agencies or sophisticated investors) is needed and should be used instead. For practical applications the adequate choice of asset default parameters and present values is of high importance. Due to our expert judgements regarding these parameters, the results of our analyses are exemplary only and cannot be linked to the bank from which the balance sheet data was taken.

© Springer Fachmedien Wiesbaden GmbH, part of Springer Nature 2018
M. Spangler, *Modelling German Covered Bonds*, Mathematische Optimierung und Wirtschaftsmathematik | Mathematical Optimization and Economathematics, https://doi.org/10.1007/978-3-658-23915-2_5

5.1.1 Nominals and Maturities

Given the finite time setting $\mathcal{T}_d := \{t_0, ..., t_S\}$ with $S \in \mathbb{N}$, $t_0 = 0$, $t_{i+1} := t_i + \Delta$ for $i = 0, ..., S - 1$ and $\Delta := \frac{0.5}{k}$, $k \in \mathbb{N}$, cf. Section 4, we set the time horizon to $t_S := T_{\max}$, i.e. to the maturity of the longest Pfandbrief outstanding at time t_0, which implies $S = \frac{t_S}{\Delta} = \frac{k \cdot T_{\max}}{0.5}$. We further assume that the risk profile of the bank's risky assets and outstanding liabilities is fully described by $4 \cdot S$ risky zero coupon bonds with nominals N_i^x and maturities $T_i^x = t_i$, where $i = 1, ..., S$ and $x \in \{CPS, OA, PB, OL\}$, i.e. it holds that $n_x = S$. The risk-free cash account with nominal N^{CPL}, which is immediately due and accrues at the risk-free rate, is assigned to the maturity bucket t_0, $T^{CPL} = 0$. Table 5.1 summarizes the above assumptions regarding the bank's maturity profile.

Table 5.1: The bank's maturity profile.

Type	t_0	t_1	t_2	...	t_{S-1}	t_S
CPS	—	N_1^{CPS}	N_2^{CPS}	...	N_{S-1}^{CPS}	N_S^{CPS}
CPL	N^{CPL}	—	—	...	—	—
OA	—	N_1^{OA}	N_2^{OA}	...	N_{S-1}^{OA}	N_S^{OA}
PB	—	N_1^{PB}	N_2^{PB}	...	N_{S-1}^{PB}	N_S^{PB}
OL	—	N_1^{OL}	N_2^{OL}	...	N_{S-1}^{OL}	N_S^{OL}

In the following, we describe how the nominals N^{CPL} and N_i^x for $i = 1, ..., S$ and $x \in \{CPS, OA, PB, OL\}$ can be derived.

Pfandbrief and cover pool nominals. In the bank's §28 Pfandbrief statistics, the maturity structure of outstanding Pfandbriefe (PB) and the cover pool (CP) is reported in nine maturity buckets $\tilde{b}_1, ..., \tilde{b}_9$ with nominals $\tilde{N}_1^y, ..., \tilde{N}_9^y$ for $y \in \{PB, CP\}$, cf. Table 5.2.

Table 5.2: Maturity structure of the bank's outstanding Pfandbriefe (PB) and the cover pool (CP) as published in the §28 Pfandbrief statistics, $y \in \{PB, CP\}$.

Bucket	Nominal	Maturity T (in years)
$\tilde{b}_1$	$\tilde{N}_1^y$	$0 \leq T \leq 0.5$
$\tilde{b}_2$	$\tilde{N}_2^y$	$0.5 < T \leq 1$
$\tilde{b}_3$	$\tilde{N}_3^y$	$1 < T \leq 1.5$
$\tilde{b}_4$	$\tilde{N}_4^y$	$1.5 < T \leq 2$
$\tilde{b}_5$	$\tilde{N}_5^y$	$2 < T \leq 3$
$\tilde{b}_6$	$\tilde{N}_6^y$	$3 < T \leq 4$
$\tilde{b}_7$	$\tilde{N}_7^y$	$4 < T \leq 5$
$\tilde{b}_8$	$\tilde{N}_8^y$	$5 < T \leq 10$
$\tilde{b}_9$	$\tilde{N}_9^y$	$10 < T \leq \infty$

In a first step, the published nominals $\tilde{N}_1^y, ..., \tilde{N}_9^y$ are redistributed to the simulation time buckets t_i with nominals

$$
\hat{N}_i^y := \begin{cases}
\frac{\tilde{N}_1^y}{|\{t_j | 0 \leq t_j \leq 0.5\}|}, & \text{if } 0 \leq t_i \leq 0.5, \\
\frac{\tilde{N}_2^y}{|\{t_j | 0.5 < t_j \leq 1\}|}, & \text{if } 0.5 < t_i \leq 1, \\
\frac{\tilde{N}_3^y}{|\{t_j | 1 < t_j \leq 1.5\}|}, & \text{if } 1 < t_i \leq 1.5, \\
\frac{\tilde{N}_4^y}{|\{t_j | 1.5 < t_j \leq 2\}|}, & \text{if } 1.5 < t_i \leq 2, \\
\frac{\tilde{N}_5^y}{|\{t_j | 2 < t_j \leq 3\}|}, & \text{if } 2 < t_i \leq 3, \\
\frac{\tilde{N}_6^y}{|\{t_j | 3 < t_j \leq 4\}|}, & \text{if } 3 < t_i \leq 4, \\
\frac{\tilde{N}_7^y}{|\{t_j | 4 < t_j \leq 5\}|}, & \text{if } 4 < t_i \leq 5, \\
\frac{\tilde{N}_8^y}{|\{t_j | 5 < t_j \leq 10\}|}, & \text{if } 5 < t_i \leq 10, \\
\frac{\tilde{N}_9^y}{|\{t_j | t_j > 10\}|}, & \text{if } t_i > 10,
\end{cases}
$$

for $y \in \{PB, CP\}$ and $i, j = 1, ..., S$. The nominals $\hat{N}_i^y$ are then adjusted for coupons by compounding the associated coupon payments to time t_i and adding these positions to the original nominals,

$$
N_i^y := \hat{N}_i^y + c_y \, \Delta \, \hat{N}_i^y \cdot \sum_{j=1}^{i} \frac{d_{*b}(0, t_j)}{d_{*b}(0, t_i)},
$$

for $y \in \{PB, CP\}$ and $i = 1, ..., S$, with the discount factor $d_{*b}(0, t)$ as defined in Observation 4.19. The coupon c_{PB} is the same for each Pfandbrief and represents on average the bank's outstanding Pfandbriefe. Similarly, c_{CP} can be interpreted as average cover pool coupon. Both coupons are determined such that the net present values $\tilde{V}_{*b}^{PB}$ and $\tilde{V}_{*b}^{CP}$ of outstanding Pfandbriefe and the cover pool as published in the bank's §28 Pfandbrief statistics are matched,

$$\sum_{i=1}^{S} \hat{N}_i^y \cdot \left(c_y \Delta \sum_{j=1}^{i} d_{*b}(0, t_j) + d_{*b}(0, t_i) \right) \overset{!}{=} \tilde{V}_{*b}^y,$$

for $y \in \{PB, CP\}$, which implies

$$c_y \overset{!}{=} \frac{\tilde{V}_{*b}^y - \sum_{i=1}^{S} \hat{N}_i^y \cdot d_{*b}(0, t_i)}{\Delta \sum_{i=1}^{S} \hat{N}_i^y \sum_{j=1}^{i} d_{*b}(0, t_j)}$$

and guarantees $\sum_{i=1}^{S} N_i^y \cdot d_{*b}(0, t_i) = \tilde{V}_{*b}^y$. To ensure that the 180-day liquidity buffer is not breached, we add an amount N^{CPL},

$$N^{CPL} := \max_{j, \, t_j \leq 0.5} \left\{ \max \left(0; \sum_{i=1}^{j} N_i^{PB} - \sum_{i=1}^{j} N_i^{CP} \right) \right\},$$

for $j = 1, ..., S$, to the cash account and reduce the cover pool positions which are not relevant for the calculation of the 180-day liquidity buffer pro rata,

$$N_i^{CPS} := \begin{cases} N_i^{CP}, & \text{if } 0 \leq t_i \leq 0.5, \\ N_i^{CP} \cdot \left(1 - \frac{N^{CPL}}{\sum_{k, \, t_k > 0.5} N_k^{CP} \cdot d_{*b}(0, t_k)} \right), & \text{if } t_i > 0.5, \end{cases}$$

for $i, k = 1, ..., S$, which requires $\sum_{k, \, t_k > 0.5} N_k^{CP} \cdot d_{*b}(0, t_k) > N^{CPL}$.[1] It then holds that

$$\sum_{i=1}^{S} N_i^{CPS} \cdot d_{*b}(0, t_i) + N^{CPL} = \tilde{V}_{*b}^{CP}.$$

---

[1] The information regarding further cover assets as published in the bank's §28 Pfand-brief statistics is not used for the determination of our 180-day liquidity buffer as this position does not have a one-to-one correspondence in our model.

We have now determined all Pfandbrief and cover pool nominals in Table 5.1.

Setting $T_{\max} := 12.5$ and $k := 1$, which implies $\Delta = 0.5$ and $S = 25$, and applying the above described calibration procedure to all Pfandbrief banks which have published their Q4 2014 §28 Pfandbrief statistics on the vdp's website[2] reveals that in all 33 cases the resulting Pfandbrief and cover pool profiles fulfil the nominal cover requirement and the excess cover requirements.

Nominal of other assets and other liabilities. In the bank's annual reports, the maturity structure of assets (A) and liabilities (L) is typically reported in four maturity buckets $\bar{b}_1, ..., \bar{b}_4$ with nominals $\tilde{N}_1^y, ..., \tilde{N}_4^y$, $y \in \{A, L\}$, cf. Table 5.3.[3] As maturity information is not always published for all balance sheet positions, we define a fifth maturity bucket $\bar{b}_5$ for the positions for which no maturity is available and set the associated nominals to $\tilde{N}_5^A := \tilde{N}^A - \sum_{i=1}^4 \tilde{N}_i^A$ and $\tilde{N}_5^L := \tilde{N}^A - \sum_{i=1}^4 \tilde{N}_i^L - \tilde{N}^E$, with $\tilde{N}^A$ being the total asset nominal and $\tilde{N}^E$ the equity capital, both of which are also published in the bank's annual report.

In the following, we assume that the assets in the bucket $\bar{b}_5$ have the same maturity profile as the remaining assets and redistribute the nominal $\tilde{N}_5^A$ pro rata to the buckets $\bar{b}_1, ..., \bar{b}_4$. For the bank's liabilities, we make a similar

---

[2]See Aareal Bank AG [1], Bayerische Landesbank [16], Berlin-Hannoversche Hypothekenbank AG [17], Bremer Landesbank [22], Commerzbank AG [32], CO-REALCREDIT BANK AG [34], DekaBank Deutsche Girozentrale [38], Deutsche Apotheker- und Aerztebank eG [39], Deutsche Genossenschafts-Hypothekenbank AG [42], Deutsche Hypothekenbank (Actien-Gesellschaft) [43], Deutsche Kreditbank AG [44], Deutsche Pfandbriefbank AG [45], Deutsche Postbank AG [46], Dexia Kommunalbank Deutschland AG [48], Duesseldorfer Hypothekenbank AG [53], DVB Bank SE [54], Hamburger Sparkasse AG [83], HSH Nordbank AG [86], ING-DiBa AG [94], Kreissparkasse Koeln [103], Landesbank Baden-Wuerttemberg [105], Landesbank Berlin [106], Landesbank Hessen-Thueringen [107], Muenchener Hypothekenbank eG [131], M.M.Warburg & CO Hypothekenbank AG [125], Natixis Pfandbriefbank AG [132], NordLB Norddeutsche Landesbank Girozentrale [133], Saar LB [140], SEB AG [143], Sparkasse KoelnBonn [146], UniCredit Bank AG [152], VALOVIS BANK AG [153], Westdeutsche ImmobilienBank AG [165], WL BANK AG Westfaelische Landschaft Bodenkreditbank [166], Wuestenrot Bank AG [167].

[3]See, for example, Sünderhauf [148], p. 150.

Table 5.3: Maturity structure of the bank's assets (A) and liabilities (L) as published in the bank's annual reports, $y \in \{A, L\}$.

Bucket	Nominal	Maturity T (in years)
$\bar{b}_1$	$\tilde{N}_1^y$	$T \leq 0.25$
$\bar{b}_2$	$\tilde{N}_2^y$	$0.25 < T \leq 1$
$\bar{b}_3$	$\tilde{N}_3^y$	$1 < T \leq 5$
$\bar{b}_4$	$\tilde{N}_4^y$	$T > 5$

assumption. By doing so, we obtain

$$\bar{N}_i^A := \tilde{N}_i^A \cdot \frac{\tilde{N}^A}{\sum_{j=1}^4 \tilde{N}_j^A} \quad \text{and} \quad \bar{N}_i^L := \tilde{N}_i^L \cdot \frac{\tilde{N}^A - \tilde{N}^E}{\sum_{j=1}^4 \tilde{N}_j^L}$$

for $i = 1, ..., 4$, with $\sum_{i=1}^4 \bar{N}_i^A = \tilde{N}^A$ and $\sum_{i=1}^4 \bar{N}_i^L + \tilde{N}^E = \tilde{N}^A$. In a next step, we map the originally reported Pfandbrief and cover pool nominals $\tilde{N}_1^z, ..., \tilde{N}_9^z$, $z \in \{PB, CP\}$, from Table 5.2 to the buckets $\bar{b}_1, ..., \bar{b}_4$ to obtain

$$\bar{N}_1^z := \frac{1}{2}\tilde{N}_1^z,$$

$$\bar{N}_2^z := \frac{1}{2}\tilde{N}_1^z + \tilde{N}_2^z,$$

$$\bar{N}_3^z := \tilde{N}_3^z + \tilde{N}_4^z + \tilde{N}_5^z + \tilde{N}_6^z + \tilde{N}_7^z,$$

$$\bar{N}_4^z := \tilde{N}_8^z + \tilde{N}_9^z.$$

The corresponding nominals $\bar{N}_i^{OA}$ and $\bar{N}_i^{OL}$ of the bank's other assets and other liabilities can now be calculated by

$$\bar{N}_i^{OA} := \bar{N}_i^A - \bar{N}_i^{CP} \quad \text{and} \quad \bar{N}_i^{OL} := \bar{N}_i^L - \bar{N}_i^{PB}$$

for $i = 1, ..., 4$. These nominals are redistributed to the simulation time buckets t_i, $i = 1, ..., S$. For $\Delta = 0.5$ this is done by

$$\hat{N}_i^x := \begin{cases} \bar{N}_1^x + \frac{1}{3}\bar{N}_2^x, & \text{if } 0 \leq t_i \leq 0.5, \\ \frac{2}{3}\bar{N}_2^x, & \text{if } 0.5 < t_i \leq 1, \\ \frac{\bar{N}_3^x}{|\{t_j | 1 < t_j \leq 5\}|}, & \text{if } 1 < t_i \leq 5, \\ \frac{\bar{N}_4^x}{|\{t_j | t_j > 5\}|}, & \text{if } t_i > 5, \end{cases}$$

for $j = 1, ..., S$ and $x \in \{OA, OL\}$, while for $\Delta < 0.5$ we define

$$\hat{N}_i^x := \begin{cases} \frac{\bar{N}_1^x}{|\{t_j \mid 0 \leq t_j \leq 0.25\}|}, & \text{if } 0 \leq t_i \leq 0.25, \\ \frac{\bar{N}_2^x}{|\{t_j \mid 0.25 < t_j \leq 1\}|}, & \text{if } 0.25 < t_i \leq 1, \\ \frac{\bar{N}_3^x}{|\{t_j \mid 1 < t_j \leq 5\}|}, & \text{if } 1 < t_i \leq 5, \\ \frac{\bar{N}_4^x}{|\{t_j \mid t_j > 5\}|}, & \text{if } t_i > 5. \end{cases}$$

As no information is available on the net present value of the bank's other assets and its other liabilities, we assume that the scaling factors are the same as for the cover pool (in the case of other assets) and outstanding Pfandbriefe (in the case of other liabilities), i.e.

$$N_i^{OA} := s_{CP} \cdot \hat{N}_i^{OA} \quad \text{and} \quad N_i^{OL} := s_{PB} \cdot \hat{N}_i^{OL}$$

for $i = 1, ..., S$, with $s_z := (\sum_{i=1}^{S} N_i^z)/(\sum_{i=1}^{S} \hat{N}_i^z)$ for $z \in \{PB, CP\}$. This fully specifies all nominals of the bank's other assets and other liabilities as given in Table 5.1.

Equity. The residual equity position is determined such that the balance sheet equation

$$N^{CPS} + N^{CPL} + N^{OA} = N^{PB} + N^{OL} + N^{EQ} \qquad \text{(BSE)}$$

holds, with $N^{CPS} := \sum_{i=1}^{S} N_i^{CPS}$, $N^{OA} := \sum_{i=1}^{S} N_i^{OA}$, $N^{PB} := \sum_{i=1}^{S} N_i^{PB}$ and $N^{OL} := \sum_{i=1}^{S} N_i^{OL}$, i.e. we obtain

$$N^{EQ} := N^{CPS} + N^{CPL} + N^{OA} - N^{PB} - N^{OL}.$$

Example. Setting t_S to $T_{\max} := 12.5$ and $k := 1$ which implies $\Delta = 0.5$ and $S = 25$ and applying the above described calibration procedure to Münchener Hypothekenbank's balance sheet and §28 Pfandbrief statistics from Q4 2014, cf. Münchener Hypothekenbank [130], [131], results in the exemplary bank balance sheet as illustrated in Figure 5.1 if a Vasicek interest rate environment with parameters $r_0 = 0.0017$, $\sigma_r = 0.0035$, $\theta_r^Q = 0.9897$ and $\kappa_r^Q = 0.0013$ is assumed.[4] Liquid cover pool assets need to be held as the net cumulative cover pool cash inflows during the next 180 days are

---

[4]For exact figures, see Table A.1 in the appendix.

not sufficient to cover cashflows from outstanding Pfandbriefe. The bank's initial balance sheet and its implications will be discussed in more detail in Chapter 6.

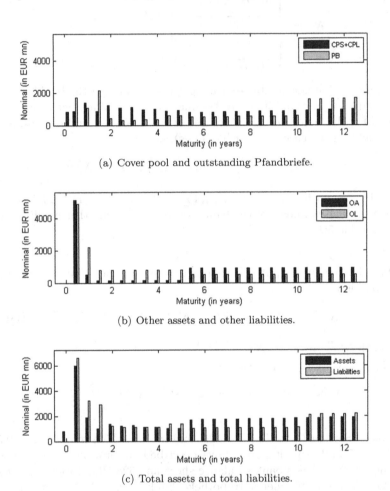

(a) Cover pool and outstanding Pfandbriefe.

(b) Other assets and other liabilities.

(c) Total assets and total liabilities.

Figure 5.1: Exemplary bank balance sheet.

5.1.2 Asset Default Parameters and Present Values

In our model, the credit performance of the bank's risky assets is driven by the assets' default parameters, their initial present values and the correlations to which the state variable processes are calibrated. For the subsequent analyses we have to set these values based on expert judgement as issuer-specific data is not publicly available. Our parameter settings are exemplary only; in practice issuer-specific data is essential and should be used instead. To get a feeling of the impact of the different parameter assumptions, sensitivity analyses will be performed in Section 6.6 below.

Default probabilities. In principle each of the modelled risky assets could be assigned its own default probability, but as information is scarce we deem it more appropriate to assign average portfolio ratings, distinguishing only between strategic cover pool assets and other assets. In the following, we assume that the bank's strategic cover pool assets all have the same rating q^{CPS} and that all other assets have the same rating q^{OA}, which is not necessarily equal to q^{CPS}. Given these ratings, the assets' lifetime default probabilities can be easily calculated from a rating migration matrix as published by rating agencies, see for example Hughes and Werner [89]. The exemplary lifetime default probabilities which are used in the following can be found in Table A.2 in the appendix. As illustrated by Figure 5.2, within a certain rating class the lifetime default probabilities are higher for longer maturities and given a fixed asset maturity they are lower for higher ratings. For a 1-year AAA bond, for example, the default probability is $\pi_{AAA}(1) = 0.01\%$, while for a 13-year AAA bond it is $\pi_{AAA}(13) = 0.42\%$. For a BB+ bond, on the other hand, we have $\pi_{BB+}(1) = 0.59\%$ and $\pi_{BB+}(13) = 16.33\%$.

Based on discussions with credit risk experts in the industry, we think that it is not unreasonable to assume q^{CPS} to lie somewhere in the BBB/BB range, depending on the actual portfolio composition and the fraction of commercial real estate assets in the cover pool. For the subsequent analyses we assume that the average cover pool rating is $q^{CPS} = \text{BB+}$. In the case of the bank's other assets, which comprise assets from non-mortgage cover pools (mostly public-sector cover pools), we also set $q^{OA} = \text{BB+}$. Figure 5.3 illustrates the resulting lifetime default probabilities of the bank's assets, with $\pi_{BB+}(0) = 0$ and linear interpolation in between integer maturity buckets. The shortest assets are assigned a lifetime default probability of $\pi_{BB+}(0.5) = 0.30\%$, while

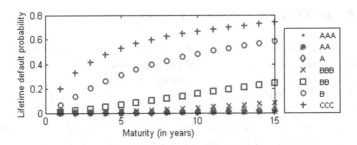

Figure 5.2: Lifetime default probabilities of selected rating classes.

the longest assets have a lifetime default probability of $\pi_{BB+}(12.5) = 15.56\%$. As already mentioned before, our choice of default parameters is exemplary only; bank-specific information is essential and should be used whenever available.

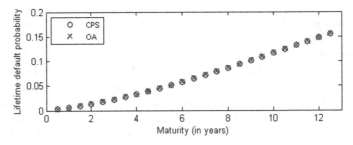

Figure 5.3: Lifetime default probabilities of the bank's risky assets.

Losses given default. With respect to the losses given default, we assume that all strategic cover pool assets have the same loss given default L^{CPS} and all other assets have the same loss given default L^{OA}, irrespective of rating or maturity. For mortgage loans, loss rates are typically lower than for other types of loans (Hagen and Holter [82]) and in the QIS 5 study the average LGD of German group 2 banks in the mortgage portfolio was found to be 27% (Deutsche Bundesbank [41]). Losses to the cover pool part of a mortgage asset do, however, only arise if foreclosure proceeds are less than

60% of the property's mortgage lending value, i.e. the loss given default of the cover pool part should be much lower than the one of the mortgage asset itself. According to Pollmann [136], the LGD of mortgage cover pool assets is considerably below 10% and Deutsche Bank [40] indicates that the LGD of first-ranking mortgage loans with loan-to-value ratios of up to 60% is about 6.3%. We therefore set $L^{CPS} = 7\%$. For the bank's other assets we assume $L^{OA} = 40\%$, which is broadly in line with average LGDs of German group 2 banks as found in the QIS 5 study.[5] Figure 5.4 illustrates the resulting expected lifetime losses of the bank's risky assets, i.e. the product of lifetime default probabilities and losses given default. For strategic cover pool assets, we get expected lifetime losses from 0.02% to 1.09% across the different asset maturities. For one year, the expected lifetime loss is 0.04%. Note that this is in line with results from a 1996 study of the institute empirica, which found loss rates for mortgage loans with loan-to value ratios of up to 60% to lie between 0.03% and 0.04%, depending on the property type (see Hagen and Holter [82]). Even though this study is clearly outdated it can still be seen as an indication that the combined choice of q^{CPS} and L^{CPS} as in our setup is not completely unrealistic. For the bank's other assets, expected lifetime losses are much higher and range from 0.12% to 6.22%.

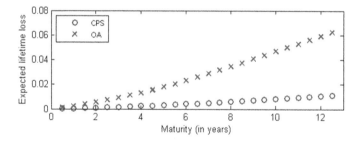

Figure 5.4: Expected lifetime losses of the bank's risky assets.

---

[5]In the QIS 5 study, average portfolio LGDs were found to be 27% in the mortgage portfolio, 37% in the corporate portfolio, 42% in the bank portfolio, 42% in the sovereign portfolio, 25% in the SME corporate portfolio, 45% in the SME retail portfolio, 51% in the other retail portfolio and 63% in the retail QRE portfolio, see Deutsche Bundesbank [41].

Asset present values. The initial spreads $s_i^x(0, T_i^x)$ of the bank's risky assets are approximated by the *inherent spreads*

$$\tilde{s}_i^x(0, T_i^x) := \frac{\tilde{\pi}_i^x \cdot \tilde{L}_i^x}{T_i^x}$$

for $i = 1, ..., n_x$ and $x \in \{CPS, OA\}$, where $\tilde{\pi}_i^x := \pi_{q^x}(T_i^x)$ is the i-th asset's lifetime default probability, $\tilde{L}_i^x = L^x$ its loss given default and T_i^x its maturity. This is of course a very rough approximation as no market risk premium is taken into account, but given the scarcity of available information and the fact that especially for illiquid mortgage assets no market prices or spread quotes exist, an approximation like this is necessary and acceptable for our purpose. Figure 5.5 shows the inherent spreads of the bank's risky assets which range from 4.1 to 8.7 bp for the bank's strategic cover pool assets and from 23.6 to 49.8 bp for the bank's other assets.

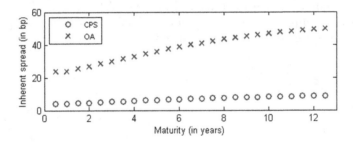

Figure 5.5: Inherent spreads of the bank's risky assets.

The assets' initial present values $\tilde{P}_i^x(0, T_i^x)$ can now be approximated by the corresponding *inherent values*

$$\tilde{I}_i^x(0, T_i^x) := e^{-(r(0, T_i^x) + \tilde{s}_i^x(0, T_i^x)) \cdot T_i^x} = e^{-(r(0, T_i^x) \cdot T_i^x + \tilde{\pi}_i^x \cdot \tilde{L}_i^x)},$$

which are lower for longer maturities, higher PDs and higher LGDs. From a pricing perspective, these are desirable features.

Correlations. The correlation matrix M of the state variable processes as specified by Assumption 9 is set based on expert judgement. Correlations between assets of the same type are assumed to be comparably high and set to $\rho^{CPS,CPS} := \rho^{OA,OA} := 0.8$, i.e. $\rho^{CPS,CPS}_{i,k} = \rho^{OA,OA}_{j,l} = 0.8$ for $i \neq k, j \neq l$, with $i, k = 1, ..., n_{CPS}$ and $j, l = 1, ..., n_{OA}$, while correlations between strategic cover pool assets and other assets are assumed to be slightly lower at $\rho^{CPS,OA} := 0.7$ due to increased diversification benefits, i.e. $\rho^{CPS,OA}_{i,j} = 0.7$. Furthermore, correlations between the state variable processes and the short rate are set to $\rho^{CPS,r}_i = \rho^{CPS,r} := -0.25$ and $\rho^{OA,r}_j = \rho^{OA,r} := -0.25$, to reflect a weak negative correlation between interest rates and real estate markets. It can be easily checked that the resulting correlation matrix M is positive definite.

Table 5.4 summarizes our parameter settings for the bank's risky assets. The associated correlation parameters can be found in Table 5.5. These parameters will be used for the calibration of the state variable processes, cf. Section 5.2.2 below.

Table 5.4: Parameter settings for the bank's risky assets, $i = 1, ..., n_{CPS}$ and $j = 1, ..., n_{OA}$.

Parameter	Setting	Value/range
$\tilde{\pi}^{CPS}_i$	$\pi_{BB+}(T^{CPS}_i)$	[0.03%; 15.56%]
$\tilde{L}^{CPS}_i$	L^{CPS}	7%
$\tilde{s}^{CPS}_i(0, T^{CPS}_i)$	$\frac{1}{T^{CPS}_i} \cdot \tilde{\pi}^{CPS}_i \cdot \tilde{L}^{CPS}_i$	[4.1bp; 8.7bp]
$\tilde{\pi}^{OA}_j$	$\pi_{BB+}(T^{OA}_j)$	[0.03%; 15.56%]
$\tilde{L}^{OA}_j$	L^{OA}	40%
$\tilde{s}^{OA}_j(0, T^{OA}_j)$	$\frac{1}{T^{OA}_j} \cdot \tilde{\pi}^{OA}_j \cdot \tilde{L}^{OA}_j$	[23.6bp; 49.8bp]

Table 5.5: Correlation settings, $i, k = 1, ..., n_{CPS}$ and $j, l = 1, ..., n_{OA}$.

Correlation	Value
$\rho^{CPS,CPS}_{i,k}, i \neq k$	0.8
$\rho^{OA,OA}_{j,l}, j \neq l$	0.8
$\rho^{CPS,OA}_{i,j}$	0.7
$\rho^{CPS,r}_i$	−0.25
$\rho^{OA,r}_j$	−0.25

5.1.3 Further Model Parameters

Apart from asset nominals, maturities, default parameters and present values which have been defined above, several other model parameters need to be set. These include the funding and liquidation haircuts and the parameters specifying the default barrier.

Funding haircuts. Funding haircuts protect the counterpart of a funding transaction from potential losses due to uncertainties regarding the future realisable sale price of pledged assets. In practice, the haircuts at which assets are pledged depend on the market environment and on asset-specific characteristics.[6] Given the lack of publicly available bank-specific asset information, we assume in the following that the funding haircuts h_{PL}^{OA} and h_{PL}^{CPS} represent the average haircut applicable to the respective asset category. For the Pfandbrief bank's other assets, which typically comprise unsecured claims against governments, credit institutions and corporates as well as secured assets like covered bonds, an average haircut of $\tilde{h}_{PL}^{OA} = 0.10$ seems plausible and is broadly in line with typical ECB haircuts.[7] As already mentioned before, the majority of Mortgage Pfandbrief issuers also have other Pfandbrief types outstanding which are not explicitly modelled in our setup. A certain fraction of other assets is therefore needed as mandatory overcollateralization for outstanding non-mortgage Pfandbriefe and reduces the amount of other assets effectively available as collateral for funding activities. In line with Assumption 30b, which states that assets in excess of the nominal cover requirement only remain in the mortgage cover pool as long as they are not needed for funding purposes, we assume that under funding pressure only the minimum required overcollateralization is maintained for non-mortgage cover pools. In the following, the fraction of other assets needed as mandatory overcollateralization for the bank's outstanding non-mortgage Pfandbriefe is therefore approximated by $p_e := \frac{N^{NM}}{N^{OA}}$, where N^{OA} is the total nominal of the bank's other assets and N^{NM} the nominal of the

---

[6] In Eurosystem monetary policy operations valuation, haircuts are assigned based on asset credit quality, liquidity category, residual maturity and coupon type, see King and Will [101]. As a rule of thumb, the lower the liquidity or the credit quality and the longer the residual maturity, the higher is this haircut.

[7] According to King and Will [101], the average haircut for coupon bonds in the liquidity categories I to IV with credit quality step 1/2 and 3 and residual maturities of up to 10 years is 0.11.

bank's outstanding non-mortgage Pfandbriefe. Effectively, a haircut of h_{PL}^{OA} has to be applied to the bank's other assets such that

$$(1 - h_{PL}^{OA}) \overset{!}{=} (1 - \tilde{h}_{PL}^{OA}) \cdot (1 - p_e).$$

Solving for h_{PL}^{OA} yields

$$h_{PL}^{OA} := 1 - (1 - p_e) \cdot (1 - \tilde{h}_{PL}^{OA}). \tag{5.1}$$

The above approximation implicitly assumes that the relative size of non-mortgage Pfandbriefe to the bank's other assets remains constant over time which is not necessarily fulfilled in practice. For banks with considerable amounts of non-mortgage Pfandbriefe, an explicit modelling of these positions might be considered. Related model extensions are, however, beyond the scope of this work. Applying the formula in Equation (5.1) to Münchener Hypothekenbank's balance sheet and its §28 Pfandbrief statistics as of Q4 2014, cf. Münchener Hypothekenbank [130], [131], we get $p_e = 33.6\%$ and with $\tilde{h}_{PL}^{OA} = 0.10$ it follows that $h_{PL}^{OA} = 40.2\%$. For strategic cover pool assets, we set $h_{PL}^{CPS} = 0.25$ in the following, to account for the fact that mortgage cover pools typically contain rather illiquid assets.

Liquidation haircuts. Similar to funding haircuts, liquidation haircuts refer to a stressed market environment and represent some kind of downturn expectation. We thus set $h_{LI}^{OA} = h_{PL}^{OA} =: h^{OA}$ and $h_{LI}^{CPS} = h_{PL}^{CPS} =: h^{CPS}$.

Default barrier. The parameters M_{ST} and M_{LT}, which define the contribution of future debt payments to the solvency barriers $B_B^O(t)$ and $B_P^O(t)$, are set to $M_{ST} = 1$ and $M_{LT} = 0.5$, i.e. short term debt is fully accounted for, while long term debt is included with a factor of 0.5 only. In the following, we consider a time period of up to 0.5 years as short term ($T_{ST} = 0.5$) and payments beyond two and a half years are considered to be long term ($T_{LT} = 2.5$). As we will see in Section 6.3, the exact choice of these parameters is less crucial in our setting as there is a second default reason (illiquidity) which mitigates the risk of underestimating default probabilities.

Table 5.6 summarizes our parameter settings for further model parameters.

Table 5.6: Further model parameter settings.

Parameter	h_{PL}^{CPS}	h_{PL}^{OA}	h_{LI}^{CPS}	h_{LI}^{OA}	M_{ST}	M_{LT}	T_{ST}	T_{LT}
Value	0.25	0.402	0.25	0.402	1	0.5	0.5	2.5

5.2 Calibration of the Market Environment

5.2.1 Risk-Free Interest Rates

In the following, we describe how the parameters of the Vasicek dynamics in Equations (4.4) and (4.5) are determined. As the application of our simulation model is mainly real-world[8], we start with the calibration of the real-world parameters κ_r^P, θ_r^P, σ_r and r_0; the risk-neutral parameters κ_r^Q and θ_r^Q are derived in a second step.

$\mathbb{P}$-dynamics. For the calibration of the parameters κ_r^P, θ_r^P, σ_r and r_0 we need historical observations $r(t_0), ..., r(t_N)$ of the short rate. As the short rate itself is not observable in the market, we use weekly Wednesday observations of the six-month Euribor[9] from 07/01/2004 to 31/12/2014 as a proxy and consider the historical observations $\tilde{r}(t_i)|\tilde{r}(t_{i-1})$, $i = 1, ..., N$, with $t_i - t_{i-1} = \frac{7}{365} =: \Delta_r$ and $N = 573$ to be independent realizations of the short rate $r(t_i)|r(t_{i-1})$. Since the as of date of our simulation is 31/12/2014, the estimate for the short rate in t_0 is set to $\hat{r}_0 := \tilde{r}(t_N) = 0.0017$.

The parameters κ_r^P, θ_r^P and σ_r are estimated by the maximum likelihood method, where the optimization problem

$$\arg \max_{\kappa_r^P, \theta_r^P, \sigma_r > 0} LL_3(\kappa_r^P, \theta_r^P, \sigma_r) \tag{5.2}$$

is solved, with $LL_3(\kappa_r^P, \theta_r^P, \sigma_r)$ being the associated log-likelihood function which is derived as follows. According to Proposition 4.1, the short rate is

---

[8]Our main goal is to determine (real-world) default parameters for Pfandbriefe.
[9]Bloomberg Ticker: EUR006M Index.

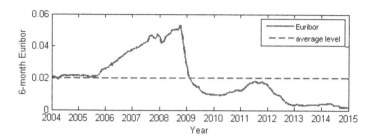

Figure 5.6: Weekly observations of the six-month Euribor from 07/01/2004 to 31/12/2014.

normally distributed and the density of $\tilde{r}(t_i)|\tilde{r}(t_{i-1})$ is given by

$$f_{\tilde{r}(t_i)|\tilde{r}(t_{i-1})}(x; \kappa_r^P, \theta_r^P, \sigma_r, \Delta_r) := \frac{1}{\hat{\sigma}_{\Delta_r}\sqrt{2\pi}} \cdot e^{-\frac{1}{2}\left(\frac{x-\hat{\mu}_{\Delta_r,i}}{\hat{\sigma}_{\Delta_r}}\right)^2}, \qquad i = 1,...,N,$$

with

$$\hat{\mu}_{\Delta_r,i} := \tilde{r}(t_{i-1})e^{-\kappa_r^P\Delta_r} + \theta_r^P(1 - e^{-\kappa_r^P\Delta_r}),$$

$$\hat{\sigma}_{\Delta_r} := \sqrt{\frac{\sigma_r^2}{2\kappa_r^P}(1 - e^{-2\kappa_r^P\Delta_r})},$$

for positive parameters κ_r^P, θ_r^P and σ_r. Since the realizations $\tilde{r}(t_i)|\tilde{r}(t_{i-1})$ are independent we get

$$LL_3(\kappa_r^P, \theta_r^P, \sigma_r) = \ln\left(\prod_{i=1}^{N} f_{\tilde{r}(t_i)|\tilde{r}(t_{i-1})}\left(\tilde{r}(t_i)|\tilde{r}(t_{i-1}); \kappa_r^P, \theta_r^P, \sigma_r, \Delta_r\right)\right)$$

$$= \sum_{i=1}^{N} \ln\left(f_{\tilde{r}(t_i)|\tilde{r}(t_{i-1})}(\tilde{r}(t_i)|\tilde{r}(t_{i-1}); \kappa_r^P, \theta_r^P, \sigma_r, \Delta_r)\right)$$

$$= -\frac{N}{2}\ln(2\pi) - \frac{N}{2}\ln\left(\frac{\sigma_r^2}{2\kappa_r^P}\left(1 - e^{-2\kappa_r^P\Delta_r}\right)\right)$$

$$- \frac{\kappa_r^P w}{\sigma_r^2(1 - e^{-2\kappa_r^P\Delta_r})},$$

where $w := \sum_{i=1}^{N} v_i^2$ and $v_i := \tilde{r}(t_i) - \tilde{r}(t_{i-1})e^{-\kappa_r^P\Delta_r} - \theta_r^P(1 - e^{-\kappa_r^P\Delta_r})$ for

$i = 1, ..., N$. For σ_r, the first-order necessary condition of the optimization problem in (5.2) results in

$$\frac{\partial LL_3(\kappa_r^P, \theta_r^P, \sigma_r)}{\partial \sigma_r} = -\frac{N}{\sigma_r} + \frac{2\kappa_r^P w}{\sigma_r^3(1 - e^{-2\kappa_r^P \Delta_r})} \overset{!}{=} 0.$$

Given that κ_r^P, Δ_r and N are strictly positive, solving for positive σ_r yields

$$\sigma_r = \sqrt{\frac{2\kappa_r^P w}{N\left(1 - e^{-2\kappa_r^P \Delta_r}\right)}} \tag{5.3}$$

if $w > 0$. Otherwise, there is no σ_r which fulfils the first-order necessary condition and LL_3 does not have a stationary point at all. In the case where $w > 0$, Equation (5.3) implies that σ_r is a function of the parameters κ_r^P and θ_r^P, and the three-dimensional optimization problem in (5.2) simplifies to

$$\arg\max_{\kappa_r^P, \theta_r^P > 0} LL_2(\kappa_r^P, \theta_r^P), \tag{5.4}$$

with

$$LL_2(\kappa_r^P, \theta_r^P) := -\frac{N}{2}\ln(2\pi) - \frac{N}{2}\ln\left(\frac{w}{N}\right) - \frac{N}{2}.$$

The first-order necessary conditions of the optimization problem in (5.4) then result in

$$\frac{\partial LL_2(\kappa_r^P, \theta_r^P)}{\partial \kappa_r^P} = -\frac{N}{w}\Delta_r\, e^{-\kappa_r^P \Delta_r}\sum_{i=1}^{N} v_i\left(\tilde{r}(t_{i-1}) - \theta_r^P\right)$$

$$= -\frac{N}{w}\Delta_r\, e^{-\kappa_r^P \Delta_r}\sum_{i=1}^{N} z_i \overset{!}{=} 0,$$

$$\frac{\partial LL_2(\kappa_r^P, \theta_r^P)}{\partial \theta_r^P} = \frac{N}{w}(1 - e^{-\kappa_r^P \Delta_r})\sum_{i=1}^{N} v_i \overset{!}{=} 0,$$

with $z_i := (\tilde{r}(t_i) - \theta_r^P) \cdot (\tilde{r}(t_{i-1}) - \theta_r^P) - e^{-\kappa_r^P \Delta_r}(\tilde{r}(t_{i-1}) - \theta_r^P)^2$ and v_i as before. Solving the first of the two above equations for κ_r^P and the second one for θ_r^P yields

$$\kappa_r^P = -\frac{1}{\Delta_r}\ln\left(\frac{\sum_{i=1}^{N}\left(\tilde{r}(t_i) - \theta_r^P\right) \cdot \left(\tilde{r}(t_{i-1}) - \theta_r^P\right)}{\sum_{i=1}^{N}(\tilde{r}(t_{i-1}) - \theta_r^P)^2}\right), \tag{5.5}$$

$$\theta_r^P = \frac{\sum_{i=1}^{N} \left(\tilde{r}(t_i) - \tilde{r}(t_{i-1})e^{-\kappa_r^P \Delta_r} \right)}{N \left(1 - e^{-\kappa_r^P \Delta_r} \right)}. \tag{5.6}$$

From Equation (5.5) it then follows that

$$e^{-\kappa_r^P \Delta_r} = \frac{\tilde{r}_{xy} - \theta_r^P \tilde{r}_x - \theta_r^P \tilde{r}_y + N(\theta_r^P)^2}{\tilde{r}_{xx} - 2\theta_r^P \tilde{r}_x + N(\theta_r^P)^2}, \tag{5.7}$$

with $\tilde{r}_y, \tilde{r}_x, \tilde{r}_{xy}$ and $\tilde{r}_{xx}$ being defined by $\tilde{r}_y := \sum_{i=1}^{N} \tilde{r}(t_i), \tilde{r}_x := \sum_{i=1}^{N} \tilde{r}(t_{i-1}),$ $\tilde{r}_{xy} := \sum_{i=1}^{N} \tilde{r}(t_i)\tilde{r}(t_{i-1})$ and $\tilde{r}_{xx} := \sum_{i=1}^{N} \tilde{r}(t_{i-1})^2.$ Substituting (5.7) in (5.6) yields

$$
\begin{aligned}
N\theta_r^P &= \frac{\tilde{r}_y - \tilde{r}_x \cdot \frac{\tilde{r}_{xy} - \theta_r^P \tilde{r}_x - \theta_r^P \tilde{r}_y + N(\theta_r^P)^2}{\tilde{r}_{xx} - 2\theta_r^P \tilde{r}_x + N(\theta_r^P)^2}}{1 - \frac{\tilde{r}_{xy} - \theta_r^P \tilde{r}_x - \theta_r^P \tilde{r}_y + N(\theta_r^P)^2}{\tilde{r}_{xx} - 2\theta_r^P \tilde{r}_x + N(\theta_r^P)^2}} \\
&= \frac{\tilde{r}_y \left(\tilde{r}_{xx} - 2\theta_r^P \tilde{r}_x + N(\theta_r^P)^2 \right) - \tilde{r}_x \left(\tilde{r}_{xy} - \theta_r^P \tilde{r}_x - \theta_r^P \tilde{r}_y + N(\theta_r^P)^2 \right)}{\left(\tilde{r}_{xx} - 2\theta_r^P \tilde{r}_x + N(\theta_r^P)^2 \right) - \left(\tilde{r}_{xy} - \theta_r^P \tilde{r}_x - \theta_r^P \tilde{r}_y + N(\theta_r^P)^2 \right)} \\
&= \frac{(\tilde{r}_y \tilde{r}_{xx} - \tilde{r}_x \tilde{r}_{xy}) + \left(\tilde{r}_x^2 - \tilde{r}_y \tilde{r}_x \right)\theta_r^P + (\tilde{r}_y - \tilde{r}_x)N(\theta_r^P)^2}{(\tilde{r}_{xx} - \tilde{r}_{xy}) + (\tilde{r}_y - \tilde{r}_x)\theta_r^P},
\end{aligned}
$$

and moving all expressions containing θ_r^P to the left results in

$$N\theta_r^P \left(\tilde{r}_{xx} - \tilde{r}_{xy} \right) - \left(\tilde{r}_x^2 - \tilde{r}_y \tilde{r}_x \right) \theta_r^P = \left(\tilde{r}_y \tilde{r}_{xx} - \tilde{r}_x \tilde{r}_{xy} \right).$$

The maximum likelihood estimate for the parameter θ_r^P is then obtained by solving for θ_r^P:

$$\hat{\theta}_r^P := \frac{\tilde{r}_y \tilde{r}_{xx} - \tilde{r}_x \tilde{r}_{xy}}{N \left(\tilde{r}_{xx} - \tilde{r}_{xy} \right) - \left(\tilde{r}_x^2 - \tilde{r}_x \tilde{r}_y \right)}.$$

The estimates for the other two parameters follow from Equations (5.3) and (5.5),

$$\hat{\kappa}_r^P := -\frac{1}{\Delta_r} \ln \left(\frac{\tilde{r}_{xy} - \tilde{r}_y \hat{\theta}_r^P - \tilde{r}_x \hat{\theta}_r^P + N(\hat{\theta}_r^P)^2}{\tilde{r}_{xx} - 2\tilde{r}_x \hat{\theta}_r^P + N(\hat{\theta}_r^P)^2} \right),$$

$$\hat{\sigma}_r^2 := \frac{2\hat{\kappa}_r^P}{N \left(1 - b^2 \right)} \left(\tilde{r}_{yy} - 2b\tilde{r}_{xy} + b^2 \tilde{r}_{xx} - d_{xy} + N(\hat{\theta}_r^P)^2 (1 - b)^2 \right),$$

with $\tilde{r}_{yy} := \sum_{i=1}^{N} \tilde{r}(t_i)^2$, $b := e^{-\hat{\kappa}_r^P \Delta_r}$ and $d_{xy} := 2(1 - b)(\tilde{r}_y - b\tilde{r}_x)\hat{\theta}_r^P.$

Applying the above formulas to the weekly observations of the six-month Euribor, we obtain $\hat{\theta}_r^P = 0.0606 =: \hat{\theta}_{r,1}^P$ and $\hat{\kappa}_r^P = -0.0432 =: \hat{\kappa}_{r,1}^P$, which is not a feasible solution of the optimization problem in (5.4) as $\hat{\kappa}_{r,1}^P$ is negative. Since the function LL_2 is well-defined and differentiable on $\mathbb{R}^+ \times \mathbb{R}^+$, it can be concluded that there is no (local) maximum on $\mathbb{R}^+ \times \mathbb{R}^+$.

Running the optimization problem in (5.4) in Matlab[10] with starting values $\theta_{r,start}^P = \kappa_{r,start}^P = 1$ while requiring that $\theta_r^P, \kappa_r^P \geq 10^{-10}$, we get $\hat{\theta}_{r,2}^P = 10^{-10}$ and $\hat{\kappa}_{r,2}^P = 0.0433$ as optimal values, i.e. the maximum is obtained for θ_r^P on the boundary.[11] As θ_r^P represents the long term average of the short rate, setting it to almost zero seems counterintuitive as the average level of short rates in our data set is considerably higher, $\bar{r} := \frac{1}{N+1} \sum_{i=0}^{N} \tilde{r}(t_i) = 0.0199$. In our case, expert judgement which does not only account for historical observations but also takes into consideration expectations about future developments seems to be more appropriate when determining the long term mean of the short rate. This is why we set $\hat{\theta}_r^P := \bar{r} = 0.0199$ in the following.

Having fixed the parameter estimate for θ_r^P, the optimization problem in (5.4) simplifies to

$$\arg \max_{\kappa_r^P > 0} LL_1(\kappa_r^P), \qquad (5.8)$$

with $LL_1(\kappa_r^P) := LL_2(\kappa_r^P, \bar{r})$. The first-order necessary condition for κ_r^P is then obtained from (5.5) with $\theta_r^P = \bar{r}$ and results in an optimal value of $\hat{\kappa}_{r,3}^P = -0.0429$, which again is not a feasible solution of the optimization problem in Equation (5.8) as it is negative. Since the function LL_1 is well-defined and differentiable on $\mathbb{R}^+$, it can be concluded that there is no (local) maximum on $\mathbb{R}^+$. Running the optimization problem in (5.8) in Matlab[12] with starting value $\kappa_{r,start}^P = 1$ while requiring $\kappa_r^P \geq 10^{-10}$, we get $\hat{\kappa}_{r,4}^P = 10^{-10}$. As this estimate for κ_r^P is almost zero, the drift term $\kappa_r^P \left(\theta_r^P - r(t) \right) dt$ which pushes $r(t)$ back to its long term mean vanishes, i.e. the mean reversion feature is lost. As can be seen from Figure 5.7, the function LL_1 is very flat on the interval $[0, 0.1]$, i.e. whether the parameter estimate for κ_r^P is set to $\hat{\kappa}_{r,4}^P$ or to a higher value as for example 0.01 or 0.05

---

[10]The Matlab function fmincon, which finds the minimum of constrained nonlinear multivariable functions, is applied to the negative of the function LL_2.

[11]When running the optimization problem in Matlab with the same starting values but requiring $\theta_r^P, \kappa_r^P \geq 0$ instead and allowing the parameters to assume a value of zero, we get $\hat{\theta}_{r,2}^P = 0$ and $\hat{\kappa}_{r,2}^P = 0.0433$, i.e. the parameter estimate for θ_r^P converges to 0.

[12]As before, the Matlab function fmincon is applied to the negative of the function which is to be maximized.

does not have much impact on the resulting value of the function LL_1. To guarantee a certain minimum level of mean reversion in our model, we claim $\hat{\kappa}_r^P := \hat{\kappa}_{r,5}^P := 0.01$, which implies that the short rate needs approximately $1/\hat{\kappa}_r^P = 100$ years to return to its long term mean.

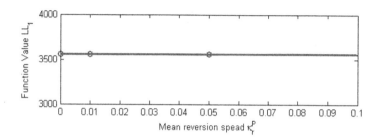

Figure 5.7: Values of the function LL_1 for $\kappa_r^P \in [0, 0.1]$.

Table 5.7 shows the feasible parameter estimates $\hat{\kappa}_{r,i}^P$ and $\hat{\theta}_{r,i}^P$, $i = 2, 4, 5$, as derived above and compares the resulting estimates $\hat{\sigma}_{r,i}$ and the associated function values $LL_2(\hat{\kappa}_{r,i}^P, \hat{\theta}_{r,i}^P)$. It turns out that the impact of the choice of the parameter set $i = 2, 4, 5$ on the value of the function LL_2 and the resulting estimate for σ_r is negligible.

Table 5.7: The feasible parameter estimates $\hat{\kappa}_{r,i}^P$, $\hat{\theta}_{r,i}^P$ and $\hat{\sigma}_{r,i}$, $i = 2, 4, 5$.

i	$\hat{\theta}_{r,i}^P$	$\hat{\kappa}_{r,i}^P$	$\hat{\sigma}_{r,i}$	LL_2
2	10^{-10}	0.0433	0.0035	3556.9
4	0.0199	10^{-10}	0.0035	3556.4
5	0.0199	0.0100	0.0035	3556.3

Figure 5.8 visualizes the two-dimensional function LL_2 for the three parameter pairs $P_i := (\hat{\theta}_{r,i}^P, \hat{\kappa}_{r,i}^P)$, $i = 2, 4, 5$.

To get a feeling how well the three feasible parameter sets fit the historical observations of the six-month Euribor, we start from $\tilde{r}(t_0) = 0.0210$ (the observation on 07/01/2004) and simulate $1,000$ paths of weekly observations

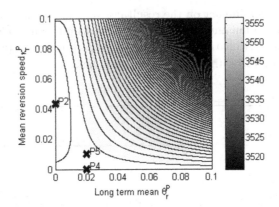

Figure 5.8: Visualization of the two-dimensional function LL_2.

for each of the tree parameter sets.[13]   The simulation results and the historical time series of the six-month Euribor are shown in Figure 5.9. Since $\tilde{r}(t_0) > \hat{\theta}^P_{r,i}$ for $i = 2, 4, 5$, the simulated paths are on average downward sloping, but for $\hat{\theta}^P_{r,2}$ the effect is more pronounced than for $\hat{\theta}^P_{r,4}$ and $\hat{\theta}^P_{r,5}$ where it is barely visible. On the other hand, the differences in the mean reversion speed which are expected to result in a lower variation of the simulated paths for higher levels of κ^P_r do not seem to be very strong. For $i = 4$ and $i = 5$ the paths appear to fit the historical observations during the financial crisis slightly better than for $i = 2$.

To further verify the adequateness of the feasible parameter sets $\hat{\kappa}^P_{r,i}$, $\hat{\theta}^P_{r,i}$, $\hat{\sigma}_{r,i}$ for $i = 1, 4, 5$, we perform a Kolmogorov-Smirnov test for goodness of fit, cf. Massey [118], which is based on the maximum difference between the empirical and the hypothetical cumulative distribution function of a set of observations. More specifically, for each of the feasible parameter sets $i = 2, 4, 5$, the observed cumulative step-function $S^i_N(x) = \frac{k^i}{N}$, where k^i is

---

[13]The paths are simulated by using the exact solution of the differential equation in the Vasicek model,

$$r(t_{j+1}) = r(t_j) \cdot e^{-\kappa^P_r \Delta_r} + \theta^P_r (1 - e^{-\kappa^P_r \Delta_r}) + \epsilon_{j+1} \sqrt{\frac{\sigma^2_r}{2\kappa^P_r} \left(1 - e^{-2\kappa^P_r \Delta_r}\right)},$$

with $\Delta_r = 7/365$ and $\epsilon_{j+1} \sim$ iid $\mathcal{N}(0,1)$ for $j = 0, ..., N-1$.

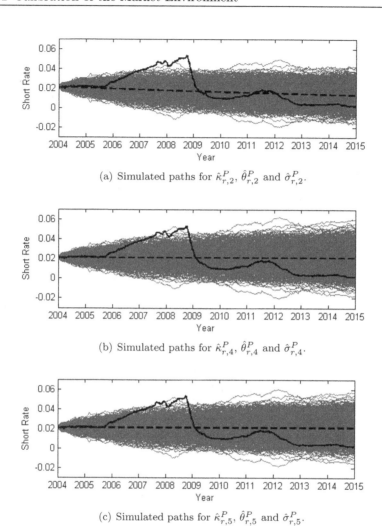

(a) Simulated paths for $\hat{\kappa}_{r,2}^P$, $\hat{\theta}_{r,2}^P$ and $\hat{\sigma}_{r,2}^P$.

(b) Simulated paths for $\hat{\kappa}_{r,4}^P$, $\hat{\theta}_{r,4}^P$ and $\hat{\sigma}_{r,4}^P$.

(c) Simulated paths for $\hat{\kappa}_{r,5}^P$, $\hat{\theta}_{r,5}^P$ and $\hat{\sigma}_{r,5}^P$.

Figure 5.9: Simulated paths (in grey), average simulated short rate level
(dashed black line) and historical observations of the six-month
Euribor (solid black line).

the number of observations y_j^i,

$$y_j^i := \frac{\tilde{r}(t_j) - \tilde{r}(t_{j-1})e^{-\hat{\kappa}_{r,i}^P \Delta_r} - \hat{\theta}_{r,i}^P(1 - e^{-\hat{\kappa}_{r,i}^P \Delta_r})}{\sqrt{\frac{(\hat{\sigma}_{r,i})^2}{2\hat{\kappa}_{r,i}^P}(1 - e^{-2\hat{\kappa}_{r,i}^P \Delta_r})}}, \quad j = 1, ..., N,$$

which are below or equal to x, is compared to the standard normal distribution function $\Phi(x)$. Under the assumptions of the Vasicek model, the observed cumulative step-function $S_N^i(x) = \frac{k^i}{N}$ is expected to be close to $\Phi(x)$. If this is not the case there is empirical evidence that the Vasicek model with parameters $\hat{\kappa}_{r,i}^P$, $\hat{\theta}_{r,i}^P$ and $\hat{\sigma}_{r,i}$ is not the correct model. The associated test statistics are given by $d_i := \max_{y \in \tilde{Y}} |\Phi(y) - S_N^i(y)|$ with $\tilde{Y} := \{y_1^i, ..., y_N^i\}$. For our sample size $N = 573$, the critical values d_α for the significance levels $\alpha = 0.05$ and $\alpha = 0.01$ are $\frac{1.36}{\sqrt{N}}$ and $\frac{1.63}{\sqrt{N}}$ respectively. If $d_i > d_\alpha$, the hypothesis that the true distribution of $S_N^i(x)$ is $\Phi(x)$ is rejected at the significance level α. Table 5.8 shows the test results for the three parameter sets and the two confidence levels. As $d_i > d_\alpha$ for $i = 2, 4, 5$ and both α values, the hypothesis that the empirical observations y_j^i, $j = 1, ..., N$, are in line with the theoretical distribution $\Phi(x)$ is rejected in all cases. This indicates that the historical time series of the six-month Euribor does not reflect the behaviour of a mean-reversion process as assumed by the Vasicek model, which makes calibration difficult in general. An investigation regarding more suitable models in this context is, however, beyond the scope of this work and left for future research. For a discussion on the pros and cons of the Vasicek model, see Section 4.3.1.

Table 5.8: Results of the Kolmogorov-Smirnov test.

i	α	d_i	d_α	reject?
2	0.01	0.1704	0.0681	yes
4	0.01	0.1630	0.0681	yes
5	0.01	0.1625	0.0681	yes
2	0.05	0.1704	0.0568	yes
4	0.05	0.1630	0.0568	yes
5	0.05	0.1625	0.0568	yes

As the Kolmogorov-Smirnov test does not give any indication that one of the two parameter sets 2 or 4 is more suitable than the parameter set 5 (d_5

is even smaller than d_2 and d_4), we set the real-world Vasicek parameters to the expert judgements in parameter set 5 and obtain

$$\hat{\theta}_r^P = 0.0199, \quad \hat{\kappa}_r^P = 0.01, \quad \hat{\sigma}_r = 0.0035, \quad \hat{r}_0 = 0.0017. \tag{RW}$$

Figure 5.10 compares the historical volatility of the weekly absolute short rate changes over a 1-year moving time window with the obtained model volatility $\hat{\sigma}_M := \sqrt{\frac{\hat{\sigma}_r^2}{2\hat{\kappa}_r^P}\left(1 - e^{-2\kappa_r^P\Delta_r}\right)}$ which results from the parameter estimates in (RW). The fit is quite good on average.

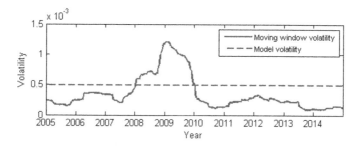

Figure 5.10: Comparison of moving window volatility and model volatility $\hat{\sigma}_M$.

$\mathbb{Q}$-dynamics. The risk-neutral short rate parameters κ_r^Q and θ_r^Q are calibrated to the yield curve observed in the market on 31/12/2014 as shown in Figure 5.11.[14] More specifically, the parameters κ_r^Q and θ_r^Q are determined by using the method of least squares, which minimizes the sum of the squared differences of observed (target) and modelled spot rates for a set of

---

[14]Bloomberg Tickers: EUR006M Index, EUFR0AG CMPN Curncy, EUFR0BH CMPN Curncy, EUFR0CI CMPN Curncy, EUFR0DJ CMPN Curncy, EUFR0EK CMPN Curncy, EUFR0F1 CMPN Curncy, EUFR0G1A CMPN Curncy, EUFR0H1B CMPN Curncy, EUFR0I1C CMPN Curncy, EUFR0J1D CMPN Curncy, EUFR0K1E CMPN Curncy, EUFR011F CMPN Curncy, EUSA2 BGN Curncy, EUSA3 BGN Curncy, EUSA4 BGN Curncy, EUSA5 BGN Curncy, EUSA6 BGN Curncy, EUSA7 BGN Curncy, EUSA8 BGN Curncy, EUSA9 BGN Curncy, EUSA10 BGN Curncy, EUSA11 BGN Curncy, EUSA12 BGN Curncy and EUSA15 BGN Curncy.

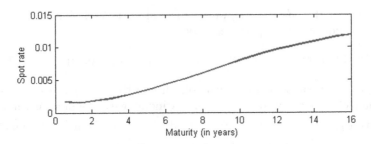

Figure 5.11: Initial yield curve as of 31/12/2014.

pre-defined yield curve buckets $\mathcal{M}$. The associated optimization problem is given by

$$\arg \min_{\kappa_r^Q, \theta_r^Q > 0} \left\{ \sum_{T \in \mathcal{M}} (\tilde{r}(t_N, t_N + T) - r(t_N, t_N + T))^2 \right\}, \qquad (5.9)$$

where $\tilde{r}(t_N, t_N + T)$ is the spot rate for a T-year investment observed in the market at time t_N and $r(t_N, t_N + T)$ is the modelled spot rate as in Equation (4.9), which is a function of the risk neutral parameters κ_r^Q and θ_r^Q and the already determined volatility σ_r. To ensure a reasonable yield curve fit, swaption volatilities are ignored. As there are no complex interest rate products to be priced in our model but only plain vanilla zero coupon bonds, a fit to swaption volatilities is of secondary importance. Furthermore, the calibration of the parameter σ_r to swaption volatilities instead of historical Euribor data would imply that long-term real-world movements of the short rate are driven by current swaption volatilities in the first place, and therefore by market risk premia, which is not desirable for our purpose.

In the following, we set $\mathcal{M} := \{0.5, 1, 2, 3, 4, 5, 6, 7, 8, 9, 10, 11, 12, 15\}$ and run the minimization problem in (5.9) in Matlab.[15] Since the minimization problem is non-linear and may have several local optima, the solution of the optimization procedure is likely to depend on the chosen pair of starting values $(\kappa_{r,start}^Q, \theta_{r,start}^Q)$. We therefore run the optimization routine 10201 times with starting values on the grid

$$\{10^{-10}, 0.01, 0.02, ..., 0.99, 1\} \times \{10^{-10}, 0.01, 0.02, ..., 0.99, 1\}$$

---

[15] Again, the Matlab function `fmincon` is used.

while requiring $\kappa_r^Q, \theta_r^Q \geq 10^{-15}$. Out of the obtained 10201 solutions, we choose the pair $(\hat{\kappa}_r^Q, \hat{\theta}_r^Q)$ which results in the lowest sum of squared differences. By doing so we obtain

$$\hat{\theta}_r^Q = 0.9897, \qquad \hat{\kappa}_r^Q = 0.0013. \qquad \text{(RN)}$$

Figure 5.12 shows the resulting yield curve fit. As of 31/12/2014, the maximum observed difference between $\tilde{r}(t_N, t_N + T)$ and $r(t_N, t_N + T)$ is 14.1 bp for $T \in \mathcal{M}$. One might argue that $\hat{\kappa}_r^Q$ and $\hat{\theta}_r^Q$ do not seem realistic in terms of interpretation[16], but we deem these parameter estimates to be acceptable as the main purpose of the risk-neutral calibration is to get the risk-free market rates right. On the other hand, the extreme realisations of $\hat{\kappa}_r^Q$ and $\hat{\theta}_r^Q$ in conjunction with our findings from the real-world calibration can be interpreted as an indication for the Vasicek model not being the most appropriate interest rate model for our purpose, especially as $\hat{\sigma}_r$ is fixed during the real-world calibration. An investigation regarding more suitable interest rate models with more degrees of freedom is, however, beyond the scope of this work and left for future research.

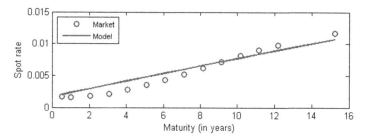

Figure 5.12: Initial yield curve fit for $\hat{\kappa}_r^Q$ and $\hat{\theta}_r^Q$.

Table 5.9 summarizes the calibrated Vasicek parameters from (RW) and (RN).

---

[16]A mean-reversion level of 99% which is reached after $1/\hat{\kappa}_r^Q = 786$ years is not intuitive from a real-world perspective.

Table 5.9: Vasicek parameter settings.

Parameter	$\hat{r}_0$	$\hat{\sigma}_r$	$\hat{\theta}_r^P$	$\hat{\kappa}_r^P$	$\hat{\kappa}_r^Q$	$\hat{\theta}_r^Q$
Value	0.0017	0.0035	0.0199	0.01	0.0013	0.9897

5.2.2 Asset Creditworthiness

The parameters of the state variable processes in Equation (4.18) are calibrated to the assets' lifetime default probabilities, their expected losses given default and their inherent spreads as specified in Section 5.1.2. In our model, the lifetime default probabilities and the losses given default of the bank's risky assets are calculated by

$$\pi_i^x := \pi_i^x(T_i^x) = \Phi\left(-\frac{\ln Z_i^{x,0} + \left(\mu_i^x - \frac{1}{2}(\sigma_i^x)^2\right) T_i^x}{\sigma_i^x \sqrt{T_i^x}}\right), \tag{5.10}$$

$$L_i^x := L_i^x(T_i^x) = 1 - \frac{Z_i^{x,0}\, e^{\mu_i^x T_i^x}}{\pi_i^x} \cdot \Phi\left(\Phi^{-1}(\pi_i^x) - \sigma_i^x \sqrt{T_i^x}\right), \tag{5.11}$$

for $i = 1, ..., n_x$ and $x \in \{CPS, OA\}$, see Lemma 4.10. Furthermore, the spreads of the risky assets are given by

$$s_i^x(0, T_i^x) = -\frac{1}{T_i^x}\left(\ln \tilde{P}_i^x(0, T_i^x) + r(0, T_i^x) T_i^x\right), \tag{5.12}$$

with $r(0, T_i^x)$ and $\tilde{P}_i^x(0, T_i^x)$ as in Equations (4.9) and (4.14).

The parameters $Z_i^{x,0}$, μ_i^x and σ_i^x of the state variable processes are then calibrated by minimizing the sum of the squared differences of the calibration targets $\tilde{\pi}_i^x$, $\tilde{L}_i^x$ and $\tilde{s}_i^x(0, T_i^x)$ from Table 5.4 and the corresponding model values π_i^x, L_i^x and $s_i^x(0, T_i^x)$ as in Equations (5.10)–(5.12) above. The associated optimization problem is given by

$$\arg\min_{Z_i^{x,0}, \sigma_i^x > 0,\ \mu_i^x \in \mathbb{R}} \left\{ LS_1^{i,x}(Z_i^{x,0}, \sigma_i^x, \mu_i^x) \right\}, \tag{5.13}$$

where

$$LS_1^{i,x}(Z_i^{x,0}, \sigma_i^x, \mu_i^x) := (\pi_i^x - \tilde{\pi}_i^x)^2 + (L_i^x - \tilde{L}_i^x)^2 + (s_i^x(0, T_i^x) - \tilde{s}_i^x(0, T_i^x))^2.$$

Solving $\pi_i^x \overset{!}{=} \tilde{\pi}_i^x$ for μ_i^x yields

$$\mu_i^x = \frac{-\Phi^{-1}(\tilde{\pi}_i^x)\,\sigma_i^x\,\sqrt{T_i^x} - \ln Z_i^{x,0} + \frac{1}{2}\,(\sigma_i^x)^2\,T_i^x}{T_i^x}, \tag{5.14}$$

i.e. for fix T_i^x, μ_i^x is a function of $Z_i^{x,0}$ and σ_i^x. The optimization problem in Equation (5.13) therefore simplifies to

$$\underset{Z_i^{x,0},\sigma_i^x > 0}{\arg\min} \left\{ LS_2^{i,x}(Z_i^{x,0},\sigma_i^x) \right\}, \tag{5.15}$$

where

$$LS_2^{i,x}(Z_i^{x,0},\sigma_i^x) := (\hat{L}_i^x - \tilde{L}_i^x)^2 + (\hat{s}_i^x(0,T_i^x) - \tilde{s}_i^x(0,T_i^x))^2.$$

The adjusted model values $\hat{L}_i^x$ and $\hat{s}_i^x(0,T_i^x)$ account for the choice of μ_i^x as in (5.14) and are given by

$$\hat{L}_i^x = 1 - \frac{Z_i^{x,0}\,e^{f_\mu(Z_i^{x,0},\sigma_i^x)\cdot T_i^x}}{\tilde{\pi}_i^x} \cdot \Phi\left(\Phi^{-1}(\tilde{\pi}_i^x) - \sigma_i^x\sqrt{T_i^x}\right),$$

$$\hat{s}_i^x(0,T_i^x) = -\frac{1}{T_i^x}\left(\ln f_{\tilde{P}}(Z_i^{x,0},\sigma_i^x) + r(0,T_i^x)\,T_i^x\right),$$

with

$$f_\mu(Z_i^{x,0},\sigma_i^x) := \frac{-\Phi^{-1}(\tilde{\pi}_i^x)\,\sigma_i^x\,\sqrt{T_i^x} - \ln Z_i^{x,0} + \frac{1}{2}\,(\sigma_i^x)^2\,T_i^x}{T_i^x},$$

$$f_{\tilde{P}}(Z_i^{x,0},\sigma_i^x) := P(0,T_i^x) - V^{\text{Put}}(0,T_i^x,1,Z_i^{x,0},\sigma_i^x,r_0,\theta_r^Q,\kappa_r^Q,\sigma_r,\rho_i^{x,r}),$$

and V^{Put} as in Equation (4.15). Note that, as $\hat{s}_i^x(0,T_i^x)$ depends on $r(0,T_i^x)$ and $f_{\tilde{P}}(Z_i^{x,0},\sigma_i^x)$, the Vasicek parameters r_0, σ_r, θ_r^Q and κ_r^Q and the correlation parameter $\rho_i^{x,r}$ have an impact on the calibration of the risky assets' parameters.

The parameter estimates $\hat{Z}_i^{x,0}$ and $\hat{\sigma}_i^x$ are then obtained by running the minimization problem in (5.15) in Matlab.[17] Having done so, the estimate

---

[17]To solve the minimization problem, the Matlab function **fminsearch** is applied to the function $LS_2^{i,x}$ with starting values $Z_{i,start}^{x,0} = \sigma_{i,start}^x = 0.1$.

for μ_i^x is derived with the help of Equation (5.14):

$$\hat{\mu}_i^x = \frac{-\Phi^{-1}(\tilde{\pi}_i^x)\,\hat{\sigma}_i^x\,\sqrt{T_i^x} - \ln \hat{Z}_i^{x,0} + \frac{1}{2}(\hat{\sigma}_i^x)^2\,T_i^x}{T_i^x}.$$

As there are sufficient degrees of freedom in the calibration, we get a perfect fit. Based on the Vasicek parameters in Table 5.9 and the correlations in Table 5.5, all observed differences $|\pi_i^x - \tilde{\pi}_i^x|$, $|L_i^x - \tilde{L}_i^x|$ and $|s_i^x(0, T_i^x) - \tilde{s}_i^x(0, T_i^x)|$ are $\leq 10^{-15}$ for all $i = 1, ..., n_x$ and $x \in \{CPS, OA\}$. The resulting model parameters $\hat{Z}_i^{x,0}$, $\hat{\mu}_i^x$ and $\hat{\sigma}_i^x$ of the bank's risky assets can be found in Table A.3 in the appendix.

5.3 Scenario Generation

Given the finite time setting $\mathcal{T}_d := \{t_0, ..., t_S\}$ with $S \in \mathbb{N}$, $t_0 = 0$, $t_{i+1} := t_i + \Delta$ for $i = 0, ..., S - 1$ and $\Delta := \frac{0.5}{k}$, $k \in \mathbb{N}$ where $t_S := T_{\max} := 12.5$, we now denote by $N \in \mathbb{N}$ the number of scenarios to be simulated. From Proposition 4.1 we know that $r(t)|\mathcal{F}_s$ is normally distributed for $t \geq s$, with mean $\mu_{r(t)|\mathcal{F}_s}$ and variance $\sigma^2_{r(t)|\mathcal{F}_s}$. A sample real-world path of the short rate can therefore be simulated by

$$r(t_{j+1}) = r(t_j)\,e^{-\kappa_r^P \Delta} + \theta_r^P(1 - e^{-\kappa_r^P \Delta}) + \epsilon_{j+1}^r \sqrt{\frac{\sigma_r^2}{2\kappa_r^P}(1 - e^{-2\kappa_r^P \Delta})},$$

for $j = 0, ..., S - 1$, with $r(t_0) := r_0$ and $\epsilon_1^r, ..., \epsilon_S^r$ being independent random draws from a standard normal distribution. Figure 5.13 shows $10,000$ simulated real-world paths of the short rate with $\Delta = 0.5$. Since $r_0 < \theta_r^P$, the mean of the short rate increases over time, but the effect is barely visible. This is not surprising as $\kappa_r^P = 0.01$ implies that it takes on average 100 years until the short rate returns to its long term mean $\theta_r^P = 0.0199$. It also becomes apparent that there is a considerable amount of negative short rates.

For the simulation of the real-world evolution of the state variable associated with the bank's i-th risky asset of type $x \in \{CPS, OA\}$, $i \in \{1, ..., n_x\}$, the analytical solution of the stochastic differential equation in Equation (4.18)

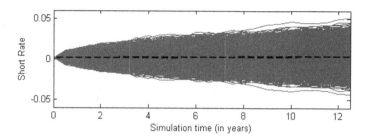

Figure 5.13: $10,000$ simulated real-world paths of the short rate (in grey) and average simulated short rate (dashed black line).

is used, cf. Brigo et al [23]:

$$Z_i^x(t_{j+1}) = Z_i^x(t_j)\, e^{(\mu_i^x - 0.5\,(\sigma_i^x)^2)\cdot\Delta + \sigma_i^x\,\sqrt{\Delta}\,\epsilon_{j+1}^{i,x}},$$

for $j = 0, ..., S - 1$, with $Z_i^x(t_0) := Z_i^{x,0}$ and $\epsilon_1^{i,x}, ..., \epsilon_S^{i,x}$ being independent random draws from a standard normal distribution. Figure 5.14 shows $10,000$ simulated real-world paths of the state variable Z_S^{CPS} which is associated with the risky cover pool asset maturing in t_S, based on $\Delta = 0.5$. The mean of the state variable paths is slightly increasing over time for this asset, which is a consequence of Remark 4.9 together with $\hat{\mu}_{25}^{CPS} = 0.0080 > 0$.

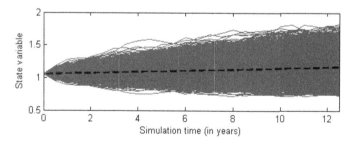

Figure 5.14: $10,000$ simulated real-world paths of the state variable Z_S^{CPS} (in grey) and average simulated state variable (dashed black line).

From the simulated short rate $r(t_{j+1})$, the price of a risk-free zero coupon bond with maturity $T \geq t_{j+1}$ is calculated according to Proposition 4.3:

$$P(t_{j+1}, T) = A(t_{j+1}, T) \cdot e^{-B(t_{j+1}, T) \cdot r(t_{j+1})}.$$

Furthermore, given the simulated state variable $Z_i^x(t_{j+1})$, the price of the associated risky zero coupon bond is obtained from Lemma 4.13, i.e. for $t_{j+1} \leq T_i^x$ we have

$$\tilde{P}(t_{j+1}, T_i^x) = P(t_{j+1}, T_i^x)$$
$$- V^{\mathrm{Put}}(t_{j+1}, T_i^x, 1, Z_i^x(t_{j+1}), \sigma_i^x, r(t_{j+1}), \theta_r^Q, \kappa_r^Q, \sigma_r, \rho_i^{x,r}).$$

To simulate N scenarios of the market environment, $N \cdot S \cdot K$ random variables need to be generated, $K := n_{CPS} + n_{OA} + 1$. For each of the variables ϵ_j^r, $\epsilon_j^{i,CPS}$ and $\epsilon_j^{k,OA}$, $i = 1, ..., n_{CPS}$ and $k = 1, ..., n_{OA}$, one random draw is needed at each of the S simulation time steps and in each of the N scenarios. The vector

$$\epsilon_j := (\epsilon_j^r, \epsilon_j^{1,CPS}, ..., \epsilon_j^{n_{CPS}, CPS}, \epsilon_j^{1,OA}, ..., \epsilon_j^{n_{OA}, OA}) \in \mathbb{R}^K$$

is normally distributed with $\epsilon_j \sim \mathcal{N}(0, M)$ for $j = 1, ..., S$ and correlation matrix M as in Assumption 9. It can be simulated by $C \tilde{\epsilon}_j$ with $\tilde{\epsilon}_j \sim \mathcal{N}(0, I_K)$, where I_K is the identity matrix of size K and C the lower triangular matrix from the Cholesky decomposition of the matrix M which fulfils $CC^\top = M$, cf. Glassermann [78], p. 82.[18] For variance reduction purposes, we use the method of antithetic variates, i.e. we effectively simulate $\tilde{N} := \frac{1}{2} N$ (with N being a multiple of 2) realizations $\epsilon_j^1, ..., \epsilon_j^{\tilde{N}}$ of the vector ϵ_j and set $\epsilon_j^{\tilde{N}+l} := -\epsilon_j^l$ for $l = 1, ..., \tilde{N}$.[19]

---

[18]In Matlab, the Cholesky decomposition of the matrix M is obtained by using the function chol.

[19]For more details on the method of antithetic variates, see Glassermann [78], Chapter 4.2.

6 Simulation Results

In the following, we discuss the simulation results obtained from the exemplary Mortgage Pfandbrief calibration as specified in the previous chapter. More specifically, we set $\Delta = 0.5$ and $T_{\max} = 12.5$, which implies $S = 25$ and $\mathcal{T}_d = \{0, 0.5, ..., 12.5\}$, and take the balance sheet profile in Table A.1 and the model parameters from Tables 5.5, 5.6, 5.9, A.3 to simulate $N = 100,000$ scenarios. Subsequently, this setup is referred to as the *base run*. The associated *certainty equivalent scenario (CE scenario)* is the deterministic scenario which is obtained when setting the volatility parameters $\hat{\sigma}_r = \hat{\sigma}_i^{CPS} = \hat{\sigma}_j^{OA} = 0$ for $i, j = 1, ..., S$. It describes the future market environment if all stochastic variables evolve "as expected", without random fluctuations. As we will see in the following, the certainty equivalent scenario can be very helpful in the context of scenario analyses. In Section 6.1 we have a first look at the default statistics from the base run and in Section 6.2 we analyse the drivers of losses in more detail. Section 6.3 is then dedicated to bank and cover pool solvency; funding and liquidity aspects are discussed in Section 6.4. In Section 6.5 we address the topics scenario quality and stability of simulation results and in Section 6.6 we perform sensitivity analyses. Finally, Section 6.7 summarizes our findings.

6.1 Default Statistics

To get a first impression of the base run results, we analyse the default statistics of the bank and the cover pool, i.e. the default probabilities (PDs), the losses given default (LGDs) and the expected losses (ELs), and look into the associated default reasons.

Default probabilities. As can be seen from Figure 6.1, the bank's default probabilities π_t^B range from 0% to 1.3%, with the first bank default occurring in year 1. In the first 3.5 years, bank defaults are still rare with a cumulative default probability of 0.05%, but they increase over time with their maximum

© Springer Fachmedien Wiesbaden GmbH, part of Springer Nature 2018
M. Spangler, *Modelling German Covered Bonds*, Mathematische Optimierung und Wirtschaftsmathematik I Mathematical Optimization and Economathematics, https://doi.org/10.1007/978-3-658-23915-2_6

being reached in year 11.5. The cumulative probability of a bank default until $T_{\max}$ is 10.4%. The cover pool default probabilities π_t^P, on the other hand, are zero prior to year 9. In later years they increase to up to 2.5% and for $t \geq 11$ they are even higher than the corresponding bank default probabilities. The cumulative probability of a cover pool default until $T_{\max}$ is 7.6%. All in all, the cover pool defaults in 73% of the bank default scenarios. Thereof, in 20% of cases the default happens at the same time as the bank default, while in 53% it occurs at some later time. In 27% of the bank default scenarios the cover pool does not default at all. Short to mid-term maturities are very safe, but longer-dated Pfandbriefe suffer from defaults and are affected by time subordination. For more detailed information on bank and cover pool default probabilities, see Tables A.4 and A.5 in the appendix.

Based on the lifetime default probabilities $\pi_R(t)$ in Table A.2 in the appendix, ratings can be assigned to the bank's liabilities. For an outstanding liability with maturity $t \in \{1, 2, ..., 13\}$, the value $\pi_R(t)$ can be interpreted as upper barrier for its cumulative default probability $\pi_{t,cum}^x$, $x \in \{B, P\}$, to be assigned a rating $R \in \{AAA, AA, A+, A-, ..., CCC, D\}$. In the case of non-integer maturities, linear interpolation is applied to obtain this upper barrier. It turns out that for Pfandbriefe with maturities of up to 10 years, the cumulative default probability of the cover pool is in line with a AAA rating. In the case of other liabilities with maturities not longer than 3.5 years, the assigned rating is also in the AAA/AA range. For longer maturities, the ratings get worse. The 12.5-year cumulative default probabilities of the bank (10.4%) and the cover pool (7.6%) correspond to a rating of BB+ and BBB- respectively. More details on the assigned PD-ratings can be found in Table A.6 in the appendix.

Losses given default and expected losses. The losses given default and the expected losses of the bank's liabilities are shown in Figure 6.2. In the case of other liabilities, the losses given default L_t^{OL} range from 15.8% to 95.2%. They are very high ($> 80\%$) most of the time but show a tendency to decrease after 10 years with the lowest non-zero loss given default being observed in $T_{\max}$. For Pfandbriefe, the losses given default L_t^{PB} are considerably lower and take values from 1.5% to 50.5% (prior to year 9 they are equal to zero), with a constant decline over time. A detailed analysis of the drivers of these losses will be performed in Section 6.2 below.

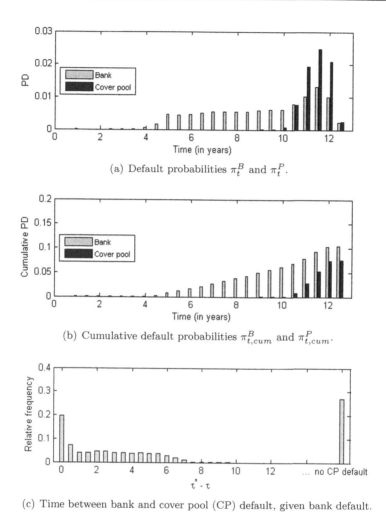

(a) Default probabilities π_t^B and π_t^P.

(b) Cumulative default probabilities $\pi_{t,cum}^B$ and $\pi_{t,cum}^P$.

(c) Time between bank and cover pool (CP) default, given bank default.

Figure 6.1: Bank and cover pool default probabilities over time.

The expected losses EL_t^{OL} of other liabilities, which are determined by multiplying the respective default probabilities and losses given default, get as high as 0.9% with a cumulative expected loss of 8.4% until $T_{\max}$. For Pfandbriefe, the expected losses EL_t^{PB} are considerably lower, especially prior to year 10, and take values of up to 0.5% with a cumulative expected loss of 1.4% until $T_{\max}$. Consequently, the average loss given default turns out to be $L_{avg}^{OL} = \frac{8.4\%}{10.4\%} = 80.5\%$ for other liabilities and $L_{avg}^{PB} = \frac{1.4\%}{7.6\%} = 19.0\%$ for Pfandbriefe. For more information on the observed losses given default and the expected losses, we refer to Table A.4 in the appendix.

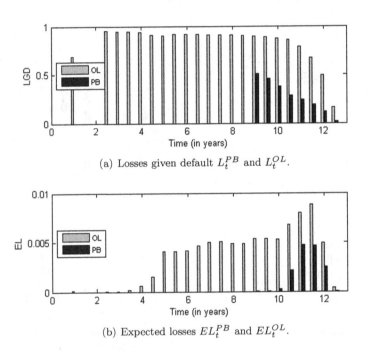

(a) Losses given default L_t^{PB} and L_t^{OL}.

(b) Expected losses EL_t^{PB} and EL_t^{OL}.

Figure 6.2: Losses given default and expected losses over time.

Default reasons. As depicted in Figure 6.3, the availability of funding plays an important role for the bank: illiquidity is the sole default reason in

61% of bank default scenarios.[1] In the first 3 years, there are only very few
bank defaults, but in all cases illiquidity is the default reason. It also remains
the predominant driver of bank default until year 10. Overindebtedness
does not seem to be a major threat for the bank in the short term. The
situation is different in the case of the cover pool which is overindebted in
the majority of cover pool default scenarios (in 88% of cases). Starting from
the first defaults in year 9, overindebtedness is always the main cover pool
default reason. For more details on the distribution of default reasons over
time, see Table A.7 in the appendix. Further in-depth analyses regarding
overindebtedness and illiquidity will be performed in Sections 6.3 and 6.4
below.

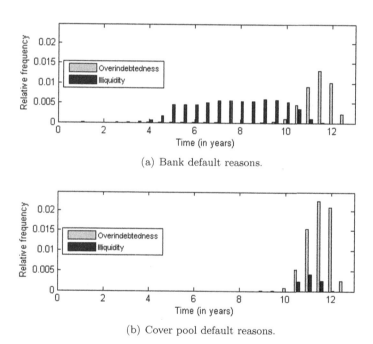

(a) Bank default reasons.

(b) Cover pool default reasons.

Figure 6.3: Relative frequency of default reasons over time (as percentage of
100, 000 scenarios).

---

[1]The bank is said to be illiquid only if it is not overindebted (Definition 4.26), but
overindebtedness does not necessarily mean that the bank is still liquid (Definition
4.25).

6.2 Analysis of Loss Drivers

In the previous section we found that losses given default vary depending on the default time. The distribution of relative losses in the single scenarios which lead to the losses given default in Figure 6.2 is shown in Figure 6.4. For other liabilities, the observed losses range from 0.2% to 100%, with an average of 80.5% and a standard deviation of 18.3%. Average losses are higher than 90% for $\tau \in [2.5, 8.5]$ and fall constantly starting from year 6.5. In case of a bank default in $T_{\max}$, the average loss is 15.8%. Pfandbrief losses are considerably lower and take values from 0.0% to 52.8%, with an average of 19.0% and a standard deviation of 7.4%. Obviously, average Pfandbrief losses decrease with increasing cover pool default time τ^*, from 50.5% in year 9 to 1.5% in $T_{\max}$. To get a better understanding of the drivers of these losses, we decompose the losses in the single scenarios into several components.

Losses of other liabilities. In the case of a bank default at time $\tau < \infty$, the loss of other liabilities is given by

$$
LS^{OL}(\tau) := \begin{cases} 1 - \frac{\hat{F}^{OL}(\tau)}{C^{OL}(\tau)}, & \text{if } \tau = \bar{\tau}, \\ 1 - \frac{\hat{F}^{OL}(\tau) + \hat{F}^{OL}(\bar{\tau}) \cdot P(\tau, \bar{\tau})}{C^{OL}(\tau)}, & \text{if } \tau < \bar{\tau}, \end{cases}
$$

$$
= 1 - k_{OL}^G(\tau) \cdot (1 - h_{LI}^{OA,avg}(\tau)) \cdot f_{OL}^V(\tau) - f_{OL}^{PA}(\tau) - f_{OL}^{CP}(\tau),
\tag{6.1}
$$

with $\bar{\tau} := \min(\tau^*, T_{\max})$. The average liquidation haircut of non-pledged other assets is denoted by $h_{LI}^{OA,avg}(\tau) := 1 - LV_{NPL}^{OA}(\tau)/V_{NPL}^{OA}(\tau)$ and the relative value of non-pledged other assets as compared to the claim of outstanding other liabilities is measured by $f_{OL}^V(\tau) := V_{NPL}^{OA}(\tau)/C^{OL}(\tau)$. Furthermore, the relative value of the subordinated claim on pledged assets is defined by $f_{OL}^{PA}(\tau) := k_{OL}^G(\tau) \cdot E^{PA}(\tau)/C^{OL}(\tau)$ and the relative value of the subordinated claim on the cover pool is given by

$$
f_{OL}^{CP}(\tau) := \begin{cases} \frac{k_{OL}^G(\tau) \cdot E^{CP}(\tau)}{C^{OL}(\tau)}, & \text{if } \tau = \bar{\tau}, \\ \frac{k_{OL}^G(\bar{\tau}) \cdot E^{CP}(\bar{\tau}) \cdot P(\tau, \bar{\tau})}{C^{OL}(\tau)}, & \text{if } \tau < \bar{\tau}. \end{cases}
$$

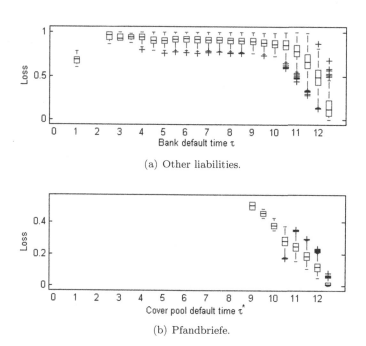

(a) Other liabilities.

(b) Pfandbriefe.

Figure 6.4: Distribution of relative loss sizes by default time (central mark of boxes: median, edges of boxes: 25th and 75th percentiles).

Obviously, the higher the values $k_{OL}^G(\tau)$, $f_{OL}^V(\tau)$, $f_{OL}^{PA}(\tau)$ and $f_{OL}^{CP}(\tau)$ and the lower $h_{LI}^{OA,avg}(\tau)$, the lower is $LS^{OL}(\tau)$.

Table 6.1 summarizes the main statistics of the loss drivers in the base run. It can be observed that $f_{OL}^V(\tau)$, $k_{OL}^G(\tau)$ and $1 - h_{LI}^{OA,avg}(\tau)$ play an important role in the context of loss sizes, with average values of 26%, 89% and 68%. As $f_{OL}^V(\tau) = V_{NPL}^{OA}(\tau)/C^{OL}(\tau) = \left(1 - k_{Pl}^{OA}(\tau - \Delta)\right) \cdot V^{OA}(\tau)/C^{OL}(\tau)$, the relative value $f_{OL}^V(\tau)$ comprises not only effects from the performance of other assets and from potential reinvestment activities but also information on previous asset pledging activities. Indeed, asset pledging turns out to have a considerable impact on losses with $k_{Pl}^{OA}(\tau - \Delta)$ being quite high on average (75%) and values reaching up to 100% in single scenarios. On the other hand, the relative value of the subordinated claims $f_{OL}^{PA}(\tau)$ and $f_{OL}^{CP}(\tau)$ is rather low on average (3% and 1% respectively) but reaches higher values in single scenarios. Other liabilities obviously suffer from structural subordination and asset encumbrance caused by secured funding activities (Pfandbrief issuance and funding from the collateralized liquidity line). This observation is not surprising and in line with expected behaviour patterns. Ahnert et al [4] also find that the issuance of secured debt asymmetrically shifts risk onto unsecured debt holders (other liabilities in our model). As can be seen from Figure A.1 in the appendix, in our example the variations in loss sizes are mainly driven by the relative value $f_{OL}^V(\tau)$ of remaining other assets, which already comprises potential changes in $k_{PL}^{OA}(\tau - \Delta)$, and by the fraction $k_{OL}^G(\tau)$.

Table 6.1: Main statistics of the loss drivers of other liabilities according to Equation (6.1).

Loss driver	Avg	Std	Max	Min	Range	Impact
$k_{OL}^G(\tau)$	89%	15%	100%	45%	55%	decreasing
$h_{LI}^{OA,avg}(\tau)$	32%	8%	40%	0%	40%	increasing
$f_{OL}^V(\tau)$	26%	26%	181%	0%	181%	decreasing
... $k_{PL}^{OA}(\tau - \Delta)$	75%	25%	100%	0%	100%	increasing
$f_{OL}^{PA}(\tau)$	3%	6%	53%	0%	53%	decreasing
$f_{OL}^{CP}(\tau)$	1%	2%	18%	0%	18%	decreasing

A closer look at the scenarios with the ten highest losses of other liabilities reveals that the top ten losses are equal to 100%.[2] In all these cases, the bank defaults quite early, between year 2.5 and year 4, with the default reason being either overindebtedness (in three cases) or illiquidity (in seven cases). The fractions $k_{OL}^G(\tau)$ are between 74% and 100%, but as $k_{PL}^{OA}(\tau - \Delta) = 100\%$ and $f_{OL}^{PA}(\tau) = 0\%$, all other assets have been pledged at the time prior to bank default and are needed to repay the bank's liquidity line at time τ, i.e. all assets are encumbered and there is nothing left in the general insolvency estate. Furthermore, the cover pool defaults at some later time $\tau^* \in [10, 12]$ in all ten scenarios, which results in $f_{OL}^{CP}(\tau) = 0$ and means that there are no excess cover pool assets which can be used to satisfy the residual claim $RC^{OL}(\tau^*)$ of other liabilities.

---

[2] Altogether, there are 12 (out of a total of $100,000$ scenarios with $10,448$ bank defaults) with a loss of 100%.

In the scenarios with the ten lowest losses of other liabilities, losses range from 0.2% to 1.1%. In all ten cases, the bank defaults in $T_{\max}$, meaning that there are no liquidation losses and $h_{LI}^{OA}(\tau) = 0$, which generally implies higher recoveries. The reason for default is always overindebtedness, with $f_{OL}^{V}(\tau) > 83\%$ and $k_{PL}^{OA}(\tau - \Delta) < 39\%$ in nine out of the ten scenarios. In the remaining scenario we have $f_{OL}^{V}(\tau) = 46\%$ and $k_{PL}^{OA}(\tau - \Delta) = 72\%$, but with $f_{OL}^{V}(\tau) = 48\%$ there are comparably high liquidation proceeds from pledged assets which are released to the general insolvency estate.[3] Furthermore, the cover pool does not default at all in five out of the ten scenarios with $f_{OL}^{CP}(\tau)$ taking values of up to 6%, i.e. cover pool assets which are not needed for the repayment of Pfandbriefe are released to the general insolvency estate. For more details on the scenarios with the ten highest and the ten lowest losses of other liabilities, see Table A.8 in the appendix.

Pfandbrief losses. In the case of a cover pool default at time $\tau^* < \infty$, Pfandbrief losses are given by

$$LS^{PB}(\tau^*) := 1 - \frac{\hat{F}^{PB}(\tau^*)}{C^{PB}(\tau^*)}$$

$$= 1 - k_{PB}^{CP}(\tau^*) \cdot (1 - h_{LI}^{CP,avg}(\tau^*)) \cdot f_{PB}^{V}(\tau^*) - f_{PB}^{GIE}(\tau^*). \tag{6.2}$$

Here, the average liquidation haircut of non-pledged cover pool assets is denoted by $h_{LI}^{CP,avg}(\tau^*) := 1 - LV_{NPL}^{CP}(\tau^*)/(V_{NPL}^{CPS}(\tau^*) + N^{CPL}(\tau^*))$ and the relative value of non-pledged cover pool assets as compared to the claim of outstanding Pfandbriefe is $f_{PB}^{V}(\tau^*) := (V_{NPL}^{CPS}(\tau^*) + N^{CPL}(\tau^*))/C^{PB}(\tau^*)$. In addition, the relative value of assets from the general insolvency estate is $f_{PB}^{GIE}(\tau^*) := k_{PB}^{G}(\tau^*) \cdot LV^{IE}(\tau^*)/C^{PB}(\tau^*)$. Obviously, the higher $k_{PB}^{CP}(\tau^*)$, $f_{PB}^{V}(\tau^*)$ and $f_{PB}^{GIE}(\tau^*)$ and the lower $h_{LI}^{CP,avg}(\tau^*)$, the lower is $LS^{PB}(\tau^*)$.

Table 6.2 summarizes the main statistics of the loss drivers in the base run. It turns out that $k_{PB}^{CP}(\tau^*)$ is 100% in all scenarios, i.e. liquidation proceeds from the cover pool do not need to be shared with the cover pool's liquidity line. Moreover, $f_{PB}^{V}(\tau^*)$ and $1 - h_{LI}^{CP,avg}(\tau^*)$ take average values of 91% and 87%. As $f_{PB}^{V}(\tau^*) = \left((1 - k_{PL}^{CPS}(\tau^* - \Delta)) \cdot V^{CPS}(\tau^*) + N^{CPL}(\tau^*)\right)/C^{PB}(\tau^*)$, the relative value $f_{PB}^{V}(\tau^*)$ does not only comprise effects from the performance of cover pool assets but also from potential previous asset pledging activities. However, as $k_{PL}^{CPS}(\tau^* - \Delta) = 0$ in all scenarios, there are never

---

[3]In the other nine scenarios, the fraction $f_{OL}^{PA}(\tau)$ is much lower with values up to 18%.

pledged cover pool assets at the time of cover pool default. The relative value $f_{PB}^V(\tau^*)$ of non-pledged cover pool assets, which implicitly also contains the Pfandbrief holder's priority claim on the cover pool, is much higher than the relative value $f_{OL}^V(\tau)$ of the claim of other liabilities on the general insolvency estate: 91% vs. 26%. In addition, the average liquidation haircut $h_{LI}^{CP,avg}(\tau^*) = 13\%$ is much lower than $h_{LI}^{OA,avg}(\tau) = 32\%$. These two observations already explain to a large degree the differences in losses given default between other liabilities and Pfandbriefe. The value of the additional claim of Pfandbrief holders on the general insolvency estate, measured by $f_{PB}^{GIE}(\tau^*)$, is comparably small in our example, with an average of 2% and a standard deviation of 3%. Further investigations reveal that, apart from the fact that we only account for this claim in the case of a simultaneous bank and cover pool default, this is caused by the low liquidation proceeds of the general insolvency estate.[4] The fraction of the claim of Pfandbrief holders at the time of cover pool default which is not satisfied by the liquidation proceeds of non-pledged cover pool assets, measured by $\frac{G^{PB}(\tau^*)}{C^{PB}(\tau^*)}$, is still 20.5% on average. Despite the low value of $f_{PB}^{GIE}(\tau^*)$ in our example, it can therefore not be concluded that the value of the additional claim of Pfandbrief holders on the general insolvency estate is negligible in general, see related discussions in Hillenbrand [84]. As can be seen from Figure A.2 in the appendix, in the base run the variations in loss sizes are solely driven by the relative value $f_{PB}^V(\tau^*)$ of the cover pool, the realized liquidation haircut $h_{LI}^{CP,avg}(\tau^*)$ and the relative value $f_{PB}^{GIE}(\tau^*)$ of assets from the general insolvency estate.

Table 6.2: Main statistics of Pfandbrief loss drivers according to Equation (6.2).

Loss driver	Avg	Std	Max	Min	Range	Impact
$k_{PB}^{CP}(\tau^*)$	100%	0%	100%	100%	0%	decreasing
$h_{LI}^{CP,avg}(\tau^*)$	13%	4%	21%	0%	21%	increasing
$f_{PB}^V(\tau^*)$	91%	6%	100%	59%	41%	decreasing
... $k_{PL}^{CPS}(\tau^* - \Delta)$	0%	0%	0%	0%	0%	increasing
$f_{PB}^{GIE}(\tau^*)$	2%	3%	27%	0%	27%	decreasing

---

[4]For the bivariate distribution of $k_{PB}^G(\tau^*)$ and $\frac{LV^{IE}(\tau^*)}{C^{PB}(\tau^*)}$, see Figure A.3 in the appendix. On average, $\frac{LV^{IE}(\tau^*)}{C^{PB}(\tau^*)}$ is only 4.4%, while $k_{PB}^G(\tau^*)$ is 25.8%.

As in the case of other liabilities, we now examine the scenarios with the ten highest and the ten lowest Pfandbrief losses. The ten highest losses range from 44.0% to 52.8% and are observed between year 9 and 9.5, with cover pool overindebtedness being the reason of default. It turns out that $f_{PB}^V(\tau^*)$ is between 59% and 68% in all cases, i.e. the value of the cover pool is quite low as compared to the claim of Pfandbrief holders. Due to the fact that the cover pool defaults prior to $T_{\max}$, a liquidation haircut of $h_{LI}^{CPS} = 25\%$ applies to outstanding strategic cover pool assets, which results in an average cover pool liquidation haircut $h_{LI}^{CP,avg}(\tau^*)$ between 16% and 20%. In addition, in all ten scenarios the bank has already defaulted at some earlier time $\tau \in [4, 5.5]$, which is why Pfandbrief holders do not have a claim against the bank's general insolvency estate, i.e. $f_{PB}^{GIE}(\tau^*) = 0$.

The ten lowest Pfandbrief losses take values between 0.001% and 0.034%. In all ten cases, the default reason is overindebtedness, but as $k_{PL}^{CPS}(\tau^* - \Delta) = 0$ and $f_{PB}^V(\tau^*) > 95\%$, the value of the cover pool is quite close to the claim of Pfandbrief holders. In addition, the cover pool defaults in $T_{\max}$, i.e. there are no liquidation losses and $h_{LI}^{CP,avg}(\tau^*) = 0$. Due to the fact that there is a simultaneous bank default in all ten scenarios, Pfandbrief holders also have a claim against the general insolvency estate, but this claim is only of comparably low value with $f_{PB}^{GIE}(\tau^*) < 5\%$. More details on the scenarios with the ten highest and the ten lowest Pfandbrief losses can be found in Table A.9 in the appendix.

A short remark on the liquidity lines. The bank's liquidity line is affected by an event of bank default in $10,058$ out of the $10,448$ bank default scenarios in the base run, i.e. in 96% of cases. In these scenarios, the ratio of the losses of the bank's liquidity line to the losses of other liabilities ranges from 0% to 32.1% and is on average 4.4% with a standard deviation of 5.4%, which implies that the losses of the bank's liquidity line are significantly below the ones of other liabilities. The claim of the liquidity line in these scenarios is on average 59% of the corresponding average claim of other liabilities. The associated bivariate loss distribution in Figure A.4 in the appendix clearly shows the loss mitigation effect of the liquidity line's priority claim on pledged assets. The losses of the bank's liquidity line are much lower than those of other liabilities. In the base run, the cover pool's liquidity line is not affected by an event of cover pool default.

6.3 Bank and Cover Pool Solvency

From Section 6.1 we know that overindebtedness plays an important role in the base run, especially in the cover pool's case. In the following, we examine the solvency situation of the bank and the cover pool and investigate the impact of the default barrier in more detail.

Bank solvency. To get a first impression of the bank's solvency situation, we compare the certainty equivalent scenario to an exemplary scenario where the bank defaults due to overindebtedness. As can be seen from Figure 6.5, in the certainty equivalent scenario in which interest rates and the assets' creditworthiness evolve "as expected", the present value $V^O_{B,CE}(t)$ of the bank's assets is above the default barrier $B^O_{B,CE}(t)$ at all times. In the exemplary bank default scenario, the bank defaults at time $\tau = 11$ when the value $V^O_B(t)$ of the bank's assets falls below the default barrier $B^O_B(t)$ for the first time.

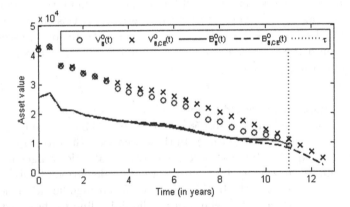

Figure 6.5: Illustration of the bank's solvency situation: certainty equivalent scenario vs. an exemplary bank default scenario.

In Section 6.1 we found that bank defaults due to overindebtedness mainly occur after the first 10 years. The bank's relative solvency buffer in the certainty equivalent scenario, which is defined by $B^S_{CE}(t) := V^O_{B,CE}(t)/B^O_{B,CE}(t) - 1$ and shown in Figure 6.6, gives an indication why. With values between

60% and 74% until year 9.5, it is comparably high in the earlier years but gets lower thereafter, taking values from 39% to 52% until year 12. This tendency of lower buffers in later years in conjunction with random fluctuations in asset present values is likely to result in more defaults due to overindebtedness. In the stochastic scenarios, the bank's solvency situation

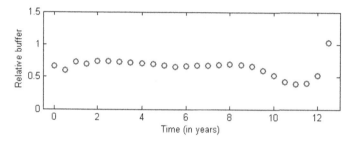

Figure 6.6: The bank's relative solvency buffer $B_{CE}^S(t)$ in the certainty equivalent scenario.

changes due to variations in the assets' present value $V_B^O(t)$ and the default barrier $B_B^O(t)$. While changes in the assets' present value are mainly driven by the market environment and potential reinvestment activities, the bank's default barrier changes due to additional liabilities from funding activities and due to discounting effects caused by changes in interest rates. From Figure A.5 in the appendix it can be seen that fluctuations in the value $V_B^O(t)$ of the bank's assets tend to be more pronounced than fluctuations in the bank's default barrier $B_B^O(t)$, especially in the earlier years. Further investigations reveal that in the scenarios where the bank defaults due to overindebtedness, the value $V_B^O(t)$ of the bank's assets at the time of default is always below $V_{B,CE}^O(t)$ and in 92% of these scenarios the barrier $B_B^O(t)$ is above $B_{B,CE}^O(t)$. It also turns out that fluctuations in asset present values are more pronounced for other assets than for the cover pool. In addition, asset nominals are rather stable except for other assets after year 10, i.e. reinvestment activities seem to be of lower importance in our example. For more details, see Figures A.6 and A.7 in the appendix.

The role of the bank's solvency situation is further investigated by a scenario analysis where we assume that bank overindebtedness does not trigger an event of bank default (Scenario O_1).[5] As illustrated by Table 6.3, the bank's cumulative default probability $\pi^B_{T_{\max},cum}$ decreases only marginally, from 10.4% to 10.3%, as the number of scenarios increases where the bank defaults due to illiquidity.

Table 6.3: Scenario analysis: the role of bank overindebtedness.

Default probability	Base run	O_1
$\pi^B_{T_{\max},cum}$	10.4%	10.3%
... thereof overindebted	4.0%	0%
... thereof illiquid	6.4%	10.3%

Cover pool solvency. The cover pool's solvency situation is illustrated by an exemplary cover pool default scenario which is shown in Figure 6.7. In this scenario, the bank is illiquid in year 8.5 while the cover pool survives until year 12 when the present value of its assets falls below the default barrier $B^O_P(t)$. As no event of bank default occurs in the certainty equivalent scenario, no information regarding the standalone cover pool's solvency situation is available for this scenario.

In the stochastic scenarios, the cover pool's solvency situation varies due to fluctuations in the cover pool's present value $V^O_P(t)$ and the default barrier $B^O_P(t)$. Similar to the bank's case, the cover pool's present value is driven by the market environment and potential reinvestment activities, while its default barrier changes due to discounting effects caused by changes in interest rates or additional liabilities from funding activities. As presented in Figure A.8 in the appendix, $V^O_P(t)$ fluctuates much more than $B^O_P(t)$ and further investigations reveal that overindebtedness is indeed mostly triggered by comparably low cover pool present values. In our example, there are never outstanding liabilities from the cover pool's liquidity line in case of a cover pool default due to overindebtedness, i.e. changes to the default barrier at this time are not triggered by funding activities. More details on

---

[5]For our scenario analysis, it is not sufficient to set the parameters M_{LT} and M_{ST} of the default barrier to 0. We need to allow for overindebtedness in $T_{\max}$ as, by definition, the bank cannot be illiquid at that time. Instead, overindebtedness in $T_{\max}$ implies that the bank is not able to fulfil its payment obligations, see Remark 4.34, which we interpret as illiquidity in the context of this scenario analysis.

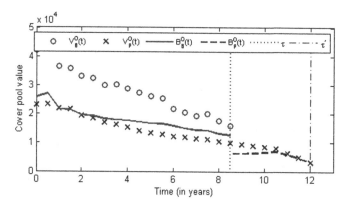

Figure 6.7: An exemplary scenario with cover pool overindebtedness.

the cover pool's solvency situation in the stochastic scenarios can be found in Figures A.9 and A.10 in the appendix.

As in the bank's case, the cover pool's solvency situation is further investigated by a scenario analysis where we assume that cover pool overindebtedness does not trigger an event of cover pool default (Scenario O_2), see Table 6.4. Again, the cumulative default probability $\pi^P_{T_{\max},cum}$ decreases only marginally, from 7.6% to 7.5% as the number of scenarios where the cover pool defaults due to illiquidity increases.

Table 6.4: Scenario analysis: the role of cover pool overindebtedness.

Default probability	Base run	O_2
$\pi^P_{T_{\max},cum}$	7.6%	7.5%
... thereof overindebted	6.7%	0%
... thereof illiquid	0.9%	7.5%

Choice of the default barrier. To get a better understanding of the role of the default barrier, we consider three alternative barrier specifications for the bank (G_1, G_2, G_3) and the cover pool (H_1, H_2, H_3) and compare them to the corresponding base run specification where $M_{ST} = 1$, $M_{LT} = T_{ST} = 0.5$

and $T_{LT} = 2.5$. The alternative default barriers G_1 and H_1 are chosen to fully account for debt of all maturities, i.e. $M_{ST} = M_{LT} = 1$ and $T_{ST} = 0$, $T_{LT} = 0.5$, while G_2 and H_2 completely ignore all debt which is not due at the respective time, $M_{ST} = 1$, $M_{LT} = T_{ST} = 0$ and $T_{LT} = 0.5$. The barriers G_3 and H_3 lie between the base run setup and G_2 and H_2, with $M_{ST} = 1$, $M_{LT} = 0$, $T_{ST} = 0.5$ and $T_{LT} = 2.5$. Figure 6.8 visualizes the different bank default barrier choices in the certainty equivalent scenario and compares them to the value $V_B^O(t)$ of the bank's assets. Obviously, fully taking into account all outstanding debt (G_1) is extremely conservative and results in the bank being close to overindebtedness already in the certainty equivalent scenario. Focusing on debt maturities at the current time (G_2) or in the near future (G_3) leaves the bank with a higher solvency buffer.

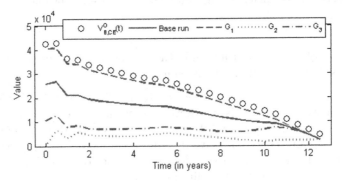

Figure 6.8: Alternative bank default barrier choices in the certainty equivalent scenario.

Table 6.5 shows the simulated default probabilities from a standalone modification of the bank and the cover pool default barrier. In the bank's case, the impact on $\pi_{T_{\max},cum}^B$ is considered, while in the cover pool's case we are interested in the impact on $\pi_{T_{\max},cum}^P$. Changing the bank's default barrier to G_1 more than doubles the bank's cumulative default probability, from 10.4% to 21.6%. Considering all outstanding debt no matter how far in the future it is due is, however, likely to be overly conservative. On the other hand, lowering the bank's default barrier to G_3 or even G_2 does not have a notable impact on $\pi_{T_{\max},cum}^B$. In the cover pool's case, the scenario H_1 increases the cumulative default probability from 7.6% to 8.5%, while H_2 and H_3 leave $\pi_{T_{\max},cum}^P$ almost unchanged.

Table 6.5: Impact of alternative default barrier choices.

Default probability	Base run	G_1	G_2	G_3	H_1	H_2	H_3
$\pi^B_{T_{\max},cum}$	10.4%	21.6%	10.3%	10.4%	—	—	—
$\pi^P_{T_{\max},cum}$	7.6%	—	—	—	8.5%	7.5%	7.6%

The observation that lowering the base run default barriers $B^O_B(t)$ and $B^O_P(t)$ only has a marginal impact on default probabilities is in line with our previous findings that overindebtedness not being a default reason only leads to a small reduction of $\pi^B_{T_{\max},cum}$ and $\pi^P_{T_{\max},cum}$ as defaults due to illiquidity increase accordingly, cf. Tables 6.3 and 6.4. The model risk of underestimating default probabilities due to an incorrect (i.e. too low) specification of the default barrier is therefore mitigated as we also consider illiquidity as a default reason. In our example, the ratio of the bank's default barrier $B^O_{B,CE}(t)$ and its outstanding debt $N^L_{CE}(t) := N^{PB}_{CE}(t) + N^{OL}_{CE}(t) + N^{LL}_{B,CE}(t)$ is between 59% and 100% in the certainty equivalent scenario, with an average of 70% over time and a tendency of higher values in later years, see Figure 6.9. Comparing these values to Davydenko [36], who uses market values of defaulting companies and finds mean default barrier estimates of 66% of the face value of outstanding debt, and to Huang and Huang [88], who assume a default boundary of around 60%, there is no indication that our default barrier is chosen too low. In later years our default barrier might even be a bit conservative, but as already mentioned before, lowering the barrier does not lead to a notable change in default probabilities.

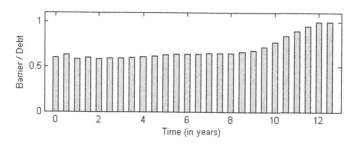

Figure 6.9: Bank default barrier $B^O_{B,CE}(t)$ vs. outstanding debt $N^L_{CE}(t)$ in the certainty equivalent scenario.

6.4 Funding and Liquidity

In Section 6.1 we found that in the base run the availability of funding plays an important role for the bank, while for the cover pool it seems to be less important. In the following, we analyse the funding situation of the bank and the cover pool in more detail and perform scenario analyses in the context of funding.

Bank funding. The bank's high dependence on funding in the stochastic scenarios is obvious from Figure 6.10, which depicts the relative frequency of a bank funding need. It turns out that from year 1 to year 9.5 the bank needs funding in all relevant scenarios where the bank has not defaulted prior to the respective time. Thereafter, the relative frequency of a bank funding need drops continuously from 39% in year 10 to 2% in year 12.[6] All in all, across all time steps prior to $T_{\max}$, the bank needs funding in 75% of cases given that it has not defaulted, but funding is not always granted when needed.

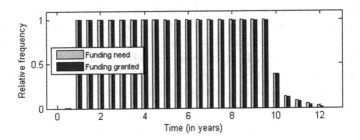

Figure 6.10: Relative frequency of a bank funding need in the base run, given that the bank has not defaulted prior to the respective time.

To get a better understanding of the bank's funding situation, we have a look at the certainty equivalent scenario. As can be seen from Figure 6.11(a), already in this scenario the bank heavily depends on funding with a funding

---

[6]In our model, the bank's funding need in $T_{\max}$ is set to 0 for technical reasons, cf. Definition 4.21. This does not necessarily mean that the bank has enough cash to fulfil its payment obligations, but in case of insufficient cash in $T_{\max}$ the bank defaults due to overindebtedness by model construction.

need at all times from year 1 to year 9.5, but there are always sufficient other assets available for pledging. Strategic cover pool assets are not available for pledging from year 4.5 to year 10. This is a result of the bank's initial balance sheet according to which the percentage nominal overcollateralization of the cover pool is negative from year 4.5 to year 12, with the gap becoming increasingly negative until year 10, see Figure 6.11(b). Consequently, the nominal cover needs to be reestablished and no excess strategic cover pool assets are available for funding activities. As depicted in Figure 6.11(c), the bank's funding need is mostly stemming from cash flow mismatches, but a certain amount of funding is also needed to reestablish cover requirements, especially in the later years. More specifically, in year 1 the funding need is caused by cash flow mismatches and the need to reestablish the 180d liquidity rule, while between year 1.5 and year 3.5 it is entirely triggered by cash flow mismatches. Thereafter, both cash flow mismatches and the need to reestablish cover requirements are responsible for the funding gaps, with the contribution of cover requirements increasing continually from year 4 to year 9.5. The relative importance of cash flow mismatches decreases accordingly. Further analyses reveal that the observed cash flow mismatches in the first years are a direct consequence of the bank's initial balance sheet which implies a negative cumulative net cash flow from year 1 to year 7, see Figure 6.11(d).[7]

Given the observations from the certainty equivalent scenario, the bank's high dependence on funding is not surprising. It also turns out that the bank's funding need and the available funding from other assets fluctuate considerably in the stochastic scenarios, with the distributions seeming rather asymmetric with unfavourable patterns, see Figure A.11 in the appendix. In year 9.5, for example, the bank's funding need can get about 9 times as high as in the certainty equivalent scenario. Similarly, the available funding from other assets may turn out to be significantly lower than in the certainty equivalent scenario. In year 12 the smallest observed value is only about $\frac{1}{10}$ of the one in the certainty equivalent scenario. Relying solely on the certainty equivalent scenario for the planning of funding activities is therefore not advisable; conditions can turn out to be much less favourable.

---

[7]It should be kept in mind that the cumulative net cash flow as implied by the bank's initial balance sheet, i.e. if there were no asset liability management activities by the bank, does not account for funding to be repaid to the bank's liquidity line or potential asset defaults. The effective cumulative net cash flow in the certainty equivalent scenario can therefore get even more negative than the one in Figure 6.11(d).

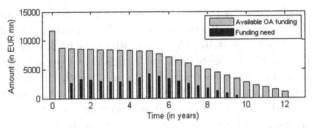

(a) Bank funding need vs. availability of funding.

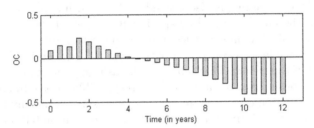

(b) Percentage nominal overcollateralization (OC).

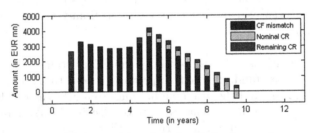

(c) Composition of the bank's funding need.

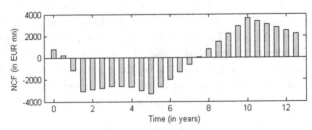

(d) Net cumulative cash flow (NCF) as implied by the bank's initial balance sheet.

Figure 6.11: The bank's funding situation in the certainty equivalent scenario.

The bank's high dependence on funding is further confirmed by scenario analyses where different degrees of funding availability are compared to the base run. In the first scenario, we do not allow for bank funding by pledging strategic cover pool assets (Scenario B_1). Then, we omit the possibility of funding by pledging other assets (Scenario B_2) and finally we disregard both funding alternatives (Scenario B_3). Table 6.6 shows the bank's cumulative default probability $\pi_{T_{\max},cum}^{B}$ and the average loss given default L_{avg}^{OL} of other liabilities in all these scenarios. Omitting bank funding by pledging strategic cover pool assets leaves $\pi_{T_{\max},cum}^{B}$ unchanged, solely the default reasons change slightly. Out of the $10,448$ bank defaults there are now 3 more defaults due to illiquidity as compared to the base run. L_{avg}^{OL} also remains almost unchanged. Omitting the possibility of bank funding by pledging other assets or bank funding in general both results in a bank default due to illiquidity with a probability of 100%. Given the bank's funding need right from the beginning, it is not surprising that most of these defaults happen already in year one 1 and some even earlier in year 0.5. The average loss given default is 27.9% in both cases and therefore much lower than in the base run. Further investigations reveal that this is mainly due to a much higher value of $f_{OL}^{V}(\tau)$ (there are no pledged assets from previous funding activities) and higher values of $k_{OL}^{G}(\tau)$ and $f_{OL}^{CP}(\tau)$, see Figure A.12 in the appendix.

Table 6.6: Scenario analysis: the role of bank funding (B_1: no bank funding by pledging strategic cover pool assets; B_2: no bank funding by pledging other assets; B_3: no bank funding at all).

Default statistics	Base run	B_1	B_2	B_3
$\pi_{T_{\max},cum}^{B}$	10.4%	10.4%	100%	100%
... thereof overindebted	4.0%	4.0%	0%	0%
... thereof illiquid	6.4%	6.4%	100%	100%
L_{avg}^{OL}	80.5%	80.4%	27.9%	27.9%

The above scenario analyses clearly demonstrate the crucial role of bank funding in the base run. The availability of other assets for pledging purposes is of high importance for our bank to survive; strategic cover pool assets on the other hand seem to only play a minor role. This is not surprising as the potential funding from strategic cover pool assets is in general much lower than the available funding from other assets. Furthermore, the nominal overcollateralization of the cover pool is 0 between year 4.5 and year 10 in all relevant scenarios, i.e. no funding from pledging strategic cover pool assets is available to the bank, which is why there is no difference between the base run and scenario B_1 at these times. For more details on the above findings, see Figure A.11 in the appendix. Note that the funding haircuts h_{PL}^{CPS} and h_{PL}^{OA} also play an important role in the context of funding. Sensitivity analyses regarding the impact of these haircuts will be performed in Section 6.6.

Cover pool funding. The relative frequency of a cover pool funding need in the stochastic scenarios is depicted in Figure 6.12. In years 1.5 and 2, there are 15 out of the 100, 000 scenarios (0.015%) in which we observe a standalone cover pool. In all these cases, the cover pool needs funding (i.e. the relative frequency of a cover pool funding need given bank default is 100%) and the funding is granted. There are also scenarios with a cover pool funding need in $t \in \{1, 2.5, 10.5, 11, 11.5, 12\}$, but the relative frequency of a cover pool funding need is much lower at these times. While funding is granted in all scenarios with a funding need in the earlier years, the cover pool never survives a funding need between years 10.5 and 12. Across all time steps prior to $T_{\max}$, funding is needed in only 4% of cases with a standalone cover pool, meaning that the cover pool's dependence on funding is comparably low in our example. The cover pool administrator's reinvestment strategy, which concentrates on liquid cover pool assets, certainly contributes to this situation as it mitigates funding pressure.

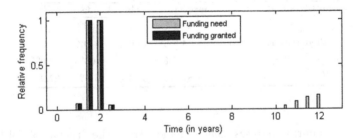

Figure 6.12: Relative frequency of a cover pool funding need in the base run, given that the bank has already defaulted but the cover pool has not defaulted prior to the respective time.

The cover pool's low dependence on funding is further confirmed by a scenario analysis where no cover pool funding is available (Scenario P_1). As presented in Table 6.7, the cumulative default probability $\pi^P_{T_{\max},cum}$ and the average loss given default L^{PB}_{avg} of Pfandbriefe are only marginally affected; there are 15 additional defaults due to illiquidity which all happen in year 1 or year 1.5.

Table 6.7: Scenario analysis: the role of cover pool funding.

Default statistics	Base run	P_1
$\pi^P_{T_{\max},cum}$	7.6%	7.6%
... thereof overindebted	6.7%	6.7%
... thereof illiquid	0.9%	0.9%
L^{PB}_{avg}	19.0%	18.9%

6.5 Scenario Quality and Stability of Results

To get a feeling how reliable our simulation results are, we analyse the quality of the $100,000$ simulated short rate and state variable paths and investigate the stability of our base run default statistics.

Scenario quality. The quality of the simulated base run scenarios can be assessed by comparing the theoretical short rate statistics $\mu_{r(t)|\mathcal{F}_0}$ and $\sigma_{r(t)|\mathcal{F}_0}$ to their realized counterparts for each time $t \in \mathcal{T}_d$. The same can be done for the default parameters π^{CPS}_i, π^{OA}_j, L^{CPS}_i and L^{OA}_j, where $i, j = 1, ..., S$. The results are summarized in Table 6.8. The quality of the simulated short rate scenarios turns out to be quite good. The maximum absolute deviation of the realized short rate mean from $\mu_{r(t)|\mathcal{F}_0}$ is below 10^{-16} and in the case of $\sigma_{r(t)|\mathcal{F}_0}$ it is below 10^{-4}. On average, the deviations are even lower. A comparison of the theoretical asset default probabilities π^{CPS}_i and π^{OA}_j to their realized counterparts shows that these values are roughly in line with deviations becoming not bigger than 25 basis points. The average (the maximum) observed deviations are 0.08% (0.25%) for strategic cover pool assets and 0.08% (0.19%) for other assets. Moreover, the average (the maximum) deviations of realized losses given defaults from the theoretical values L_i^{CPS} and L_j^{OA} are 0.09% (0.21%) in the case of strategic cover pool assets and 0.29% (1.62%) in the case of other assets. With deviations of up to 1.62%, the differences for losses given defaults of other assets are comparably high, especially for maturities up to five years.[8] As we will see in the following, this is related to the corresponding state variable process parameters.

Table 6.8: Absolute deviations of realized statistics from their theoretical counterparts.

$r(t)$	$\mu_{r(t)\|\mathcal{F}_0}$	$\sigma_{r(t)\|\mathcal{F}_0}$		CPS	π_i^{CPS}	L_i^{CPS}		OA	π_j^{OA}	L_j^{OA}
avg	$2 \cdot 10^{-17}$	$1 \cdot 10^{-5}$		avg	0.08%	0.09%		avg	0.08%	0.29%
max	$6 \cdot 10^{-17}$	$4 \cdot 10^{-5}$		max	0.25%	0.21%		max	0.19%	1.62%

A comparison of the state variable process parameters reveals that the parameters $Z_k^{OA,0}$ and σ_k^{OA} are considerably higher than their counterparts $Z_k^{CPS,0}$ and σ_k^{CPS}, with differences being more pronounced for shorter maturities, see Figure A.14 in the appendix. As can be seen from Table 6.9, the starting values $Z_k^{OA,0}$ are on average by a factor of almost 200 higher than $Z_k^{CPS,0}$ and σ_k^{OA} is about 9 times as high as σ_k^{CPS}. The parameters μ_k^{CPS} and μ_k^{OA} are in a similar order of magnitude. As the rating assumptions are the same for strategic cover pool assets and other assets ($q^{CPS} = q^{OA} = $ BB+), the parameter differences must be caused by the differences in losses given default ($L^{OA} = 40\%$ vs. $L^{CPS} = 7\%$). For fixed rating and asset maturity, more extreme parameter constellations seem to be needed to fit the higher loss given default target. This is confirmed by the results from five additional model calibrations as described in the following.

Table 6.9: Base run state variable process parameter statistics.

CPS	$Z_i^{CPS,0}$	μ_i^{CPS}	σ_i^{CPS}		OA	$Z_i^{OA,0}$	μ_i^{OA}	σ_i^{OA}
avg	1.3	0.6%	9.4%		avg	263.8	0.8%	80.3%
max	2.0	0.8%	35.1%		max	3718.1	1.1%	303.8%
min	1.1	0.3%	4.0%		min	6.0	0.3%	33.8%

---

[8]For a more detailed view on the deviations, we refer to Figure A.13 in the appendix.

To assess the sensitivity of the calibrated state variable process parameters $Z_j^{OA,0}$, μ_j^{OA} and σ_j^{OA} with respect to the choice of the calibration target L^{OA}, we leave all other model settings (including default probabilities) unchanged and perform additional calibrations for $L^{OA} \in \{10\%, 20\%, 30\%, 50\%, 60\%\}$ instead of $L^{OA} = 40\%$. Table 6.10 compares the statistics of the obtained state variable process parameters.[9] Obviously, $Z_j^{OA,0}$ and σ_j^{OA} highly depend on the choice of L^{OA} while μ_j^{OA} remains comparably stable. For $L^{OA} = 10\%$, the average starting value $Z_j^{OA,0}$ is 1.6 but gets higher with increasing calibration targets L^{OA}, reaching an average of $1.6 \cdot 10^9$ for $L^{OA} = 60\%$. Similarly, the average value σ_j^{OA} increases from 13.8% to 176.1%.

Table 6.10: State variable process parameter statistics for different choices of $L := L^{OA} \in \{10\%, ..., 60\%\}$.

$L =$ 10%	$Z_j^{OA,0}$	μ_j^{OA}	σ_j^{OA}	$L =$ 40%	$Z_j^{OA,0}$	μ_j^{OA}	σ_j^{OA}
avg	1.6	0.7%	13.8%	avg	263.8	0.8%	80.3%
max	2.9	0.9%	51.8%	max	3718.1	1.1%	303.8%
min	1.1	0.3%	5.9%	min	6.0	0.3%	33.8%
$L =$ 20%	$Z_j^{OA,0}$	μ_j^{OA}	σ_j^{OA}	$L =$ 50%	$Z_j^{OA,0}$	μ_j^{OA}	σ_j^{OA}
avg	3.7	0.7%	30.8%	avg	$5.4 \cdot 10^4$	0.7%	119.0%
max	13.3	1.1%	115.7%	max	$1.1 \cdot 10^6$	1.0%	452.1%
min	1.6	0.3%	13.1%	min	24.4	0.2%	49.7%
$L =$ 30%	$Z_j^{OA,0}$	μ_j^{OA}	σ_j^{OA}	$L =$ 60%	$Z_j^{OA,0}$	μ_j^{OA}	σ_j^{OA}
avg	16.1	0.8%	52.2%	avg	$1.6 \cdot 10^9$	0.6%	176.1%
max	121.3	1.1%	196.8%	max	$3.9 \cdot 10^{10}$	0.9%	672.4%
min	2.6	0.3%	22.1%	min	338.7	0.2%	73.0%

Running 100,000 scenarios with the same random variables (fixed seed) for each of the above L^{OA} calibrations and comparing the obtained absolute deviations of realized loss given default statistics from their theoretical

---

[9]By model construction, the calibration of the parameters $Z_i^{CPS,0}$, μ_i^{CPS} and σ_i^{CPS} is not affected by the choice of L^{OA}.

counterparts L_j^{OA} yields the results as shown in Table 6.11. Obviously, deviations get larger for higher values of L^{OA}, for which the parameters take more extreme values. These results indicate that our asset model might not be suitable for arbitrary constellations of asset default probabilities and asset losses given defaults. Default parameter constellations which force the parameters of the state variable processes into extreme values may result in instabilities in the simulations which can, if at all, only be counteracted by increasing the number of simulated scenarios.

Table 6.11: Absolute deviations of realized statistics from their theoretical counterparts L_j^{OA} for different choices of $L^{OA} \in \{10\%, ..., 60\%\}$.

L^{OA}	10%	20%	30%	40%	50%	60%
avg	0.09%	0.17%	0.23%	0.29%	0.33%	0.36%
max	0.52%	0.97%	1.34%	1.62%	1.80%	1.85%

In order to test how much the scenario quality can be improved by increasing the number of scenarios, we compare the quality of the L_j^{OA} fit to $L^{OA} = 40\%$ for four different scenario sets: S_1 (10,000 scenarios), S_2 (100,000 scenarios), S_3 (250,000 scenarios) and S_4 (500,000 scenarios), with $S_1 \subset S_2 \subset S_3 \subset S_4$ and S_2 being equal to our base run scenario set. As can be seen from Table 6.12, average deviations tend to get smaller for increasing scenario size. For 10,000 scenarios, the average deviation from L_j^{OA} is 1.12%, while for 500,000 scenarios it is 0.19% only. Similarly, the maximum deviation decreases from 5.21% to 1.46%.

Table 6.12: Absolute deviations of realized statistics from their theoretical counterparts L_j^{OA} for increasing scenario size: S_1 (10,000 scenarios), S_2 (100,000 scenarios), S_3 (250,000 scenarios) and S_4 (500,000 scenarios).

Scenario	S_1	S_2	S_3	S_4
avg	1.12%	0.29%	0.24%	0.19%
max	5.21%	1.62%	0.77%	1.46%

Stability of simulation results. To test whether the default statistics obtained from the $100,000$ base run scenarios are sufficiently stable, we vary the choice of the random seed used for scenario generation and compare the resulting default statistics. Including the base run, we compare ten scenario sets with $100,000$ scenarios each, i.e. we generate nine additional scenario sets. The results are summarized in Table 6.13. For a better comparison, the default statistics of the base run are also shown (second column). It turns out that, even though there is a certain variation in default statistics across the 10 scenario sets, the cumulative default probabilities and the average losses given default from the base scenario are very close to the averages obtained from the ten scenario sets and differ by at most 0.2%.

Table 6.13: Default statistics from ten scenario sets with $100,000$ scenarios.

Default statistics	Base Run	Avg	Std	Max	Min	Range
$\pi^B_{T_{max},cum}$	10.4%	10.6%	0.1%	10.7%	10.4%	0.2%
$\pi^P_{T_{max},cum}$	7.6%	7.7%	0.1%	7.7%	7.5%	0.2%
L^{OL}_{avg}	80.5%	80.5%	0.2%	80.8%	80.2%	0.6%
L^{PB}_{avg}	19.0%	19.1%	0.1%	19.2%	19.0%	0.2%

We conclude that the base run default statistics obtained from $100,000$ scenarios are reasonably stable and that there is no indication that they are not representative. Increasing the amount of simulated scenarios to $250,000$ or $500,000$ would of course add additional stability, but a trade-off between accurateness and run time has to be made. As the focus of this work is not on computational aspects but on the Pfandbrief model itself, $100,000$ scenarios are acceptable for illustrational purposes.

6.6 Sensitivity Analyses

As already discussed in Chapter 5, some of our parameter settings are exemplary only. To get a better understanding of the impact of different parameter choices on the resulting default statistics, we run sensitivity analyses on asset ratings, losses given default, correlations and funding and liquidation haircuts. For ease of presentation, we concentrate on the impact on the cumulative default probabilities $\pi^B_{T_{max},cum}$ and $\pi^P_{T_{max},cum}$ and the average losses given default L_{avg}^{OL} and L_{avg}^{PB}. All impacts are given in terms of a standalone shift of the respective parameter. We also show the results of a combined stress scenario where several parameters are shocked simultaneously.

Asset rating. The sensitivity analyses with respect to asset ratings are performed for $q^{CPS}, q^{OA} \in \mathcal{M}_r := \{AAA, AA, A, BBB, BB, B, CCC\}$, see Figure 6.13.[10] As intuitively expected, the bank's cumulative default probability $\pi_{T_{\max},cum}^B$ highly depends on the choice of q^{OA}. Better asset ratings result in lower default probabilities with an observed range of 61.1% (0.8% for AAA and 61.9% for CCC). The impact of q^{CPS} on $\pi_{T_{\max},cum}^B$ is similar but less pronounced with a range of 10.8% (9.0% for AAA and 19.8% for CCC). In the cover pool's case, better ratings also result in lower default probabilities, but $\pi_{T_{\max},cum}^P$ is more sensitive to q^{CPS} than to q^{OA} (ranges: 19.4% vs. 10.7%). The impact of asset ratings on L_{avg}^{OL} and L_{avg}^{PB} is ambiguous. Here it needs to be kept in mind that initial asset default probabilities do not only have an impact on asset present values and cash flows but also on bank and cover pool default timing and therefore the circumstances of default, which in turn determine the actual loss size, see discussions in Section 6.2. While L_{avg}^{PB} is more sensitive to q^{CPS} than to q^{OA}, the opposite is true for L_{avg}^{OL}. This observation is in line with our previous findings on loss drivers.[11]

Loss given default of assets. The impact of different asset loss given default parameters $L^{CPS}, L^{OA} \in \mathcal{M}_l := \{10\%, 20\%, ..., 90\%\}$ is shown in Figure 6.14.[12] Both L^{CPS} and L^{OA} turn out to have a considerable impact on the bank's and the cover pool's default statistics. The higher the loss given default of the assets, the higher the cumulative default probability of the bank and the cover pool. While $\pi_{T_{\max},cum}^B$ is more sensitive to L^{OA}, $\pi_{T_{\max},cum}^P$ is more sensitive to L^{CPS}. The observed range of cumulative bank default probabilities is 13.9% (for L^{CPS}) and 20.5% (for L^{OA}), and for the

---

[10]We explicitly do not choose a more granular grid to avoid distortions due to default probabilities of neighboured rating classes being quite close, especially for good ratings. Otherwise, it may happen that realized asset default probabilities are higher for better ratings (violated rating hierarchy) due to stochastic fluctuations, see the results from the scenario quality analyses in Section 6.5 above.

[11]In Section 6.2 we found that Pfandbrief losses are more influenced by $f_{PB}^V(\tau^*)$ than by $f_{PB}^{GIE}(\tau^*)$, while losses of other liabilities depend more on $f_{OL}^V(\tau)$ than on $f_{OL}^{CP}(\tau)$.

[12]In our model, asset loss given default parameters equal to 0% or 100% are not permitted.

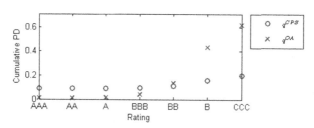

(a) Cumulative bank default probability $\pi^B_{T_{\max},cum}$.

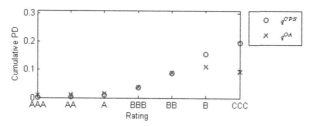

(b) Cumulative cover pool default probability $\pi^P_{T_{\max},cum}$.

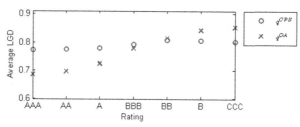

(c) Loss given default L^{OL}_{avg} of other liabilities.

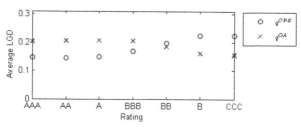

(d) Loss given default L^{PB}_{avg} of Pfandbriefe.

Figure 6.13: Rating sensitivities.

cover pool we get a range of 15.2% and 8.0% respectively. As in the case of asset ratings, the impact on losses given default is ambiguous, with loss sizes depending on the circumstances of default. L^{CPS} has a large positive impact on L_{avg}^{PB} (range: 24.1%) and a small negative impact on L_{avg}^{OL} (range: 4.7%). The opposite is true in the case of L^{OA}, which has a small negative impact on L_{avg}^{PB} (range: 2.9%) and a positive impact on L_{avg}^{OL} (range: 10.2%).

Correlations. The impact of changes in the correlation parameters is analysed for pairwise correlations in $\mathcal{M}_c := \{-100\%, -90\%, ..., 90\%, 100\%\}$ which result in a positive definite correlation matrix. In the case of $\rho^{CPS,r}$ and $\rho^{OA,r}$, this is fulfilled for $\mathcal{M}_{c_1} := \{-60\%, -50\%, ..., 10\%, 20\%\}$ and in the case of $\rho^{CPS,CPS}$ and $\rho^{OA,OA}$, it is given for $\mathcal{M}_{c_2} := \{60\%, 70\%, 80\%, 90\%\}$. For $\rho^{CPS,OA}$, positive definiteness restricts the choice of parameter values to $\mathcal{M}_{c_3} := \{-60\%, -50\%, ..., 70\%, 80\%\}$. As illustrated by Figure 6.15, the bank's and the cover pool's default statistics are not very sensitive to $\rho^{CPS,r}$ and $\rho^{OA,r}$: the range of observed values is below 1% in all eight cases. Similarly, with a maximum observed range of 2.3%, the impact of $\rho^{CPS,CPS}$ and $\rho^{OA,OA}$ is also not very big. The situation is different for $\rho^{CPS,OA}$ where $\pi_{T_{max},cum}^P$ increases from 0.5% for $\rho^{CPS,OA} = -60\%$ to 9.0% for $\rho^{CPS,OA} = 80\%$. This positive relationship intuitively makes sense as bank default is likely to be accompanied by bad asset performance. If both other assets and strategic cover pool assets perform poorly at the same time, the cover pool is more likely to default if the bank defaults. For the three other default statistics, the observed ranges of sensitivities with respect to $\rho^{CPS,OA}$ are lower (less than 6%).

Funding haircuts. The results of our sensitivity analyses with respect to funding haircuts, with $\tilde{h}_{PL}^{OA}$ and $h_{PL}^{CPS} \in \mathcal{M}_h := \{0\%, 10\%, ..., 90\%, 100\%\}$, are summarized in Figure 6.16. The bank's cumulative default probability $\pi_{T_{max},cum}^B$ turns out to be extremely sensitive to the choice of the funding haircut of other assets; the observed range of values is 89.6%. Higher haircuts $\tilde{h}_{PL}^{OA}$ result in higher default probabilities as they imply that less funding is available to the bank. For $\tilde{h}_{PL}^{OA} = 0\%$, we get a cumulative bank default probability of 10.4% until T_{max}, while for $\tilde{h}_{PL}^{OA} = 40\%$ we still have $\pi_{T_{max},cum}^B = 12.9\%$. For $\tilde{h}_{PL}^{OA} \geq 60\%$, the bank's cumulative default probability reaches 100%, i.e. the bank defaults for sure once a certain

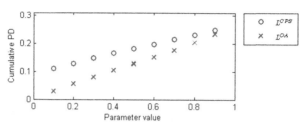

(a) Cumulative bank default probability $\pi^B_{T_{\max},cum}$.

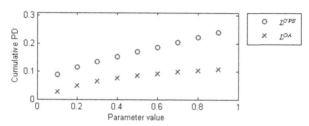

(b) Cumulative cover pool default probability $\pi^P_{T_{\max},cum}$.

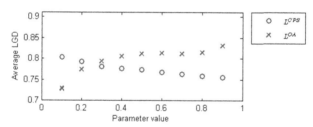

(c) Loss given default L^{OL}_{avg} of other liabilities.

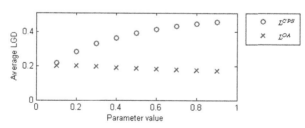

(d) Loss given default L^{PB}_{avg} of Pfandbriefe.

Figure 6.14: Loss given default sensitivities.

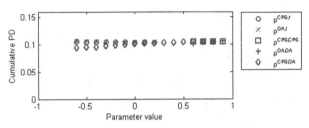

(a) Cumulative bank default probability $\pi^B_{T_{\max},cum}$.

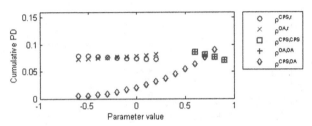

(b) Cumulative cover pool default probability $\pi^P_{T_{\max},cum}$.

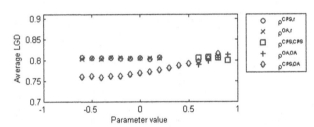

(c) Loss given default L^{OL}_{avg} of other liabilities.

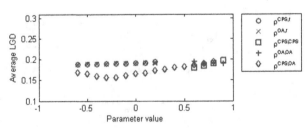

(d) Loss given default L^{PB}_{avg} of Pfandbriefe.

Figure 6.15: Correlation sensitivities.

haircut level is reached.[13] The funding haircut h_{PL}^{CPS} does not have an impact on $\pi_{T_{\max},cum}^{B}$ at all. Both findings are in line with our previous findings in the context of funding.[14] The cover pool's cumulative default probability $\pi_{T_{\max},cum}^{P}$ is also sensitive to the choice of $\tilde{h}_{PL}^{OA}$ but to a much lesser extent than $\pi_{T_{\max},cum}^{B}$. It ranges from 0.1% to 8.4% with a mostly negative impact in our example. Similar to the bank's case, the cover pool's cumulative default probability is insensitive to the choice of h_{PL}^{CPS}. As cover pool funding does not rely on asset pledging and h_{PL}^{CPS} was found to have no impact on $\pi_{T_{\max},cum}^{B}$, this does not come at a surprise. The average loss given default of other liabilities highly depends on the choice of $\tilde{h}_{PL}^{OA}$, with values ranging from 27.9% to 80.6% and a tendency of decreasing values for higher haircuts. The impact of $\tilde{h}_{PL}^{OA}$ on L_{avg}^{PB} is much lower with no clear direction and observed values between 14.4% and 19.1%. The choice of the funding haircut h_{PL}^{CPS} has no impact on L_{avg}^{OL} and L_{avg}^{PB}.

Liquidation haircuts. Figure 6.17 shows the results of our sensitivity analyses with respect to the liquidation haircuts $\tilde{h}_{LI}^{OA}$ and h_{LI}^{CPS} for values in $\mathcal{M}_h$. As intuitively expected, the cumulative default probabilities $\pi_{T_{\max},cum}^{B}$ and $\pi_{T_{\max},cum}^{P}$ are not affected by the choice of these haircuts. The average losses given defaults increase with increasing haircuts as less liquidation proceeds are available to satisfy the creditors' claims. For Pfandbriefe, the average loss given default L_{avg}^{PB} is highly sensitive to h_{LI}^{CPS}, taking values from 7.8% to 53.9%. The impact of $\tilde{h}_{LI}^{OA}$ is much lower with a range of only 1.0%. In the case of other liabilities, it is reversed: L_{avg}^{OL} is more sensitive to $\tilde{h}_{LI}^{OA}$ (range: 15.8%) than to h_{LI}^{CPS} (range: 4.2%). These findings are perfectly in line with our previous results on loss drivers from Section 6.2.[15]

---

[13] A choice of $\tilde{h}_{PL}^{OA} = 0$ corresponds to an effective funding haircut of $h_{PL}^{OA} = p_e = 33.6\%$, while $\tilde{h}_{PL}^{OA} = 60\%$ results in $h_{PL}^{OA} = 73.4\%$. As already discussed in Section 5.1.3, the effective haircut h_{PL}^{OA} takes into consideration the bank's level of asset encumbrance from other Pfandbrief types.

[14] In Section 6.4 we found that the availability of other assets for pledging purposes is of high importance for bank survival, but funding from strategic cover pool assets does not play a notable role.

[15] In Section 6.2 we found that Pfandbrief losses are, among others, driven by $h_{LI}^{CP,avg}(\tau^*)$ but not so much by $f_{PB}^{GIE}(\tau^*)$, while losses of other liabilities depend more on $h_{LI}^{OA,avg}(\tau)$ than on $f_{OL}^{CP}(\tau)$.

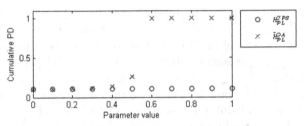

(a) Cumulative bank default probability $\pi^B_{T_{\max},cum}$.

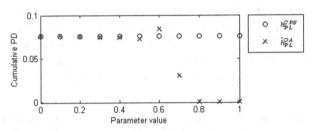

(b) Cumulative cover pool default probability $\pi^P_{T_{\max},cum}$.

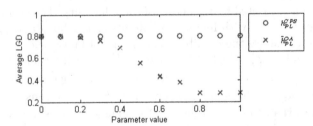

(c) Loss given default L^{OL}_{avg} of other liabilities.

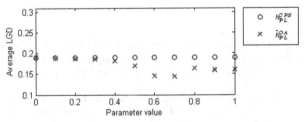

(d) Loss given default L^{PB}_{avg} of Pfandbriefe.

Figure 6.16: Sensitivities with respect to funding haircuts.

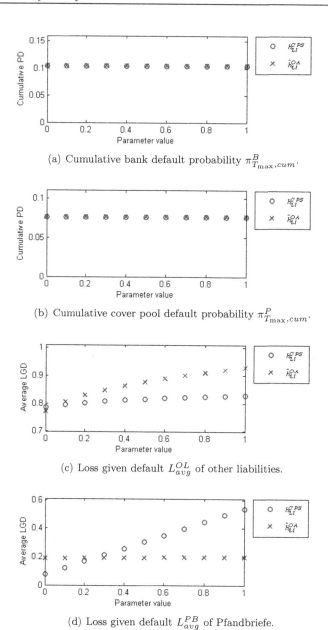

(a) Cumulative bank default probability $\pi^B_{T_{\max},cum}$.

(b) Cumulative cover pool default probability $\pi^P_{T_{\max},cum}$.

(c) Loss given default L^{OL}_{avg} of other liabilities.

(d) Loss given default L^{PB}_{avg} of Pfandbriefe.

Figure 6.17: Sensitivities with respect to liquidation haircuts.

Stress scenario. To get a feeling for the impact of combined parameter changes, we also consider a stress scenario where several parameters are shocked simultaneously in an unfavourable manner. More specifically, we set the ratings q^{CPS} and q^{OA} to BB instead of BB+, which roughly corresponds to a 4.5% (absolute) increase in the 12.5-year default probability of the risky assets. In addition, we also shock the base run losses given default L^{CPS} and L^{OA}, the funding and liquidation haircuts h_{PL}^{CPS}, $\tilde{h}_{PL}^{OA}$, h_{LI}^{CPS} and $\tilde{h}_{LI}^{OA}$ and the correlation $\rho^{CPS,OA}$ by +5% in absolute terms. The results are shown in Table 6.14. As compared to the base run, the cumulative default probabilities of the bank and the cover pool increase by 6.4% and 6.8% respectively, while the losses given default increase by 2.8% and 7.0%. In the case of $\pi_{T_{\max},cum}^{B}$ and $\pi_{T_{\max},cum}^{P}$, the impact of the combined shock (third column) is bigger than the sum of the impacts of the same but standalone stresses (last column), which indicates non-linear dependencies. For the losses given default, both impacts are roughly the same.

Table 6.14: Stress scenario results: impact of the combined stress (CS) and the sum of impacts of the standalone stresses (SS) as compared to the base run.

Default statistics	Base run	Impact of CS	Sum of impacts of SS
$\pi_{T_{\max},cum}^{B}$	10.4%	+6.4%	+5.9%
$\pi_{T_{\max},cum}^{P}$	7.6%	+6.8%	+5.4%
L_{avg}^{OL}	80.5%	+2.8%	+2.8%
L_{avg}^{PB}	19.0%	+7.0%	+7.0%

Interestingly, the increase in $\pi_{T_{\max},cum}^{B}$ in the combined stress scenario is accompanied by an increase in the default probabilities for most maturities, but the increase in $\pi_{T_{\max},cum}^{P}$ mainly affects later maturities, see Figure 6.18. Even in the combined stress scenario, Pfandbrief maturities of up to 9 years are in line with a AAA rating, with a cumulative cover pool default probability of $\pi_{T_{\max},cum}^{P}(9) = 0.18\%$. This further illustrates the high level of protection of Pfandbrief holders and also the phenomenon of time subordination.

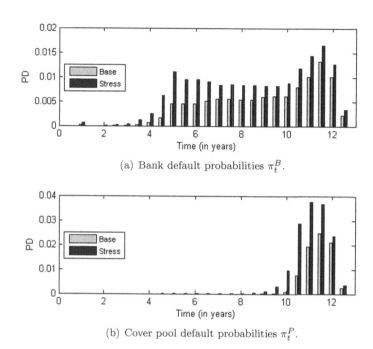

(a) Bank default probabilities π_t^B.

(b) Cover pool default probabilities π_t^P.

Figure 6.18: Stress scenario: default probabilities over time.

6.7 Summary of Simulation Results

In our base run setting, bank defaults are very rare in the first few years (AAA/AA rating), but they increase significantly over time with a cumulative default probability of 10.4% for the longest other liability (BB+ rating). Illiquidity turns out to be the main reason of bank default, especially in the short term and prior to year 10. The bank's high dependence on funding is already present in the certainty equivalent scenario where there is a bank funding need at all times between year 1 and year 9.5. This funding need is mostly caused by cash flow mismatches (and therefore a direct consequence of the bank's initial balance sheet), but to a certain extent it is also triggered by the need to reestablish cover requirements. In the certainty equivalent scenario, there are always sufficient assets available for pledging and no event of default occurs. The situation is different in the stochastic scenarios where the bank's funding need and the available funding fluctuate considerably and funding is not always granted any more. The observed patterns regarding the bank's funding need remain unchanged in the stochastic scenarios. The bank always needs funding between year 1 and year 9.5 as long as it has not defaulted prior to the respective time. The crucial role which funding plays for our bank is further demonstrated by the fact that omitting the possibility of bank funding by pledging other assets results in an almost immediate bank default. With respect to overindebtedness, we find that the bank's initial balance sheet implies a tendency of lower buffers in later years, which is clearly visible in the certainty equivalent scenario. This, in conjunction with fluctuation in asset present values and the default barrier, results in an increasing number of bank defaults due to overindebtedness in later years in the stochastic scenarios, mainly triggered by a deterioration of asset performance. To a certain degree, overindebtedness also implies illiquidity. Scenario analyses where we assume that overindebtedness does not trigger an event of default reveal no significant reduction of the bank's and the cover pool's cumulative default probabilities as defaults due to illiquidity increase accordingly. Regarding the choice of the default barrier this means that the model risk of underestimating default probabilities due to a too low default barrier choice is mitigated. Furthermore, results from the certainty equivalent scenario indicate that the ratio of the bank's default barrier and its outstanding debt is between 59% and 100%, with an average of 70% over time and a tendency of higher values in later years. This is roughly in line with values used in the literature (60% to 66%) but appears to be a bit high. In conjunction with our previous finding that lowering the default barrier does not result in a notable change in the bank's cumulative default probability, it can be concluded that there is no indication that our default barrier choice is inadequate or distorts default statistics.

Pfandbiefe with maturities of up to 10 years are quite safe in our example, and their default probabilities are in line with a AAA rating. As expected, the cumulative cover pool default probability is below the one of a bank default for all maturities. For longer-dated Pfandbriefe, there is a non-negligible risk of default with a cumulative default probability of 7.6% for the longest maturity (BBB- rating). Effectively, only in 27% of the bank default scenarios all Pfandbriefe survive. Maturities of 9 years or longer, which are affected by an event of default in our simulations, account for

46% of the initially outstanding Pfandbrief nominal. The phenomenon of a high level of protection on the one hand and the risk of potential time subordination on the other hand becomes even more visible when comparing the combined stress scenario with simultaneous adverse parameter shocks of +5% as compared to the base run. The 6.8% absolute increase in $\pi^P_{T_{\max},cum}$ mainly affects Pfandbrief maturities longer than 9 years, with shorter-dated maturities remaining in line with a AAA rating. In the case of the cover pool, overindebtedness is the main default reason, while the dependence on funding is comparably low. On the other hand, we do not find a significant reduction in the cover pool's cumulative default probability if we assume that overindebtedness does not trigger an event of cover pool default.

In the base run, the average loss given default of Pfandbriefe (19.0%) is considerably lower than the one of other liabilities (80.5%). Apart from differences in asset riskiness (default parameters of strategic cover pool assets vs. other assets) and liquidation losses as implied by the chosen liquidation haircuts, this is mainly due to the priority claim of Pfandbrief holders on the cover pool. The value of the additional claim of Pfandbrief holders on the general insolvency estate turns out to be rather small in our example. This is not solely a consequence of our assumption that this claim is only enforceable in the case of a simultaneous bank and cover pool default, but it is also caused by the low liquidation proceeds remaining in the general insolvency estate as many other assets have been pledged to the bank's liquidity line prior to default. It should not be concluded that the value of the additional claim of Pfandbrief holders on the general insolvency estate is negligible in general. The low recovery rates of other liabilities, on the other hand, are not only caused by structural subordination and the performance of other assets. They also result from the bank's high asset encumbrance from other Pfandbrief types (modelled implicitly through the liquidation haircut of other assets) and its extensive funding activities (with a high amount of assets having been pledged prior to bank default and therefore not being available to the general insolvency estate).

The results of our sensitivity analyses are plausible and reveal that the parameters specifying the creditworthiness of the bank's risky assets need to be chosen with care as they can have a large impact on the liabilities' default statistics. They do not only influence asset present values and cash flows but, more generally, bank and cover pool default timing and the circumstances of default, which in turn determine the associated losses. Similarly, the assumed funding and liquidation haircuts can also strongly influence the simulation results. The impact of correlations, on the other hand, is rather small, the only exception being the correlation between state variable processes of strategic cover pool assets and other assets. Due to non-linear dependencies, simultaneous changes in more than one of the parameters can result in a stronger effect than the sum of the effects of the standalone changes.

With respect to scenario quality and the stability of simulation results we find that the default statistics obtained from our base run with $100,000$ simulated scenarios are reasonably stable and representative. Findings from our analyses do, however, indicate that the used asset model might not be suitable for arbitrary constellations of asset default parameters. Extreme state variable process parameters obtained in the calibration can result in simulation instabilities.

All in all, given the bank's initial balance sheet and its implications on asset liability mismatches and asset riskiness, the bank highly depends on funding and there is a non-negligible probability of bank default due to illiquidity after the first few years. Providing unsecured funding for a longer time horizon would therefore not be advisable for a risk-averse investor. By contrast, Pfandbriefe are very safe for short and mid-term maturities, but longer dated Pfandbriefe are affected by time subordination. This risk could be mitigated by the introduction of additional Pfandbrief features which prevent the excessive use of cover pool proceeds for the repayment of shorter-dated Pfandbriefe, cf. Spangler and Werner [145]. As compared to other liabilities, Pfandbriefe benefit from comparably high recovery rates. This is, amongst others, due to the priority claim on the cover pool. The value of the additional claim of Pfandbrief holders on the bank's general insolvency estate, which is quite low in our example, depends on the amount of unencumbered other assets at the time of bank default. Other liabilities do not only suffer from structural subordination but also from asset encumbrance due to collateralized funding activities.

All our results do, of course, highly depend on the assumptions on which our model is based. These comprise not only the model calibration itself (e.g. the bank's initial balance sheet and the choice of our model parameters), but also more general assumptions including the choice of the default barrier and the assumption that cover pools from additional Pfandbrief types are only implicitly modelled via haircuts. Especially what happens in later years is highly influenced by our run-off consideration (e.g. no new business, no active maintenance of voluntary overcollateralization) and the assumed asset liability management (e.g. reinvestment strategies) and should therefore be considered with care. For practical applications an adequate choice of the balance sheet parameters is essential. To obtain realistic bank-specific results, more granular information on the balance sheet composition, which is typically available to the issuing bank, rating agencies or sophisticated investors, is needed and should be used instead. Due to the expert judgement on which our asset default parameters and asset present values are based, our obtained default statistics can only be considered of exemplary character.

7 Conclusion and Outlook

The purpose of this work was to develop a Pfandbrief model for structural analyses in the context of product features and risk profiles which accounts for the Pfandbrief's most important characteristics and adequately reflects its main risks. Required modelling features where derived based on the legal framework for Pfandbrief issuance and an analysis of product-specific risks. A multi-period simulation-based framework in which all terms can be easily calculated by recursive computation turned out to be most suitable for our purpose. Similar cash flow approaches are commonly used by insurance companies, for the market-consistent valuation of insurance liabilities.

Our Pfandbrief model, which is inspired by Sünderhauf [148] and Liang et al [114], is based on the distinction of nine Pfandbrief scenarios which depend on the current simulation time and the occurrence of the events of bank and cover pool default. Each of these scenarios triggers different liability cash flows and asset liability management activities. The bank's balance sheet consists of several classes of assets and liabilities, with different nominals, maturities and riskiness, and therefore allows for an adequate representation of asset liability mismatches and an explicit distinction of the balance sheet positions which are related to the bank's Pfandbrief business. To avoid an artificial change of the risk profile over time, we work in in a run-off setup with no new business being made. The market environment is represented through the stochastic dynamics of the short rate and the creditworthiness of the bank's risky assets. Due to our specific model assumptions, the impact of these risk drivers on the assets' present values and cash flows (and therefore on the bank's balance sheet) can be calculated analytically. The asset liability management in our model is dynamic and accounts for funding and reinvestment strategies as well as for the maintenance of overcollateralization according to the legal requirements. The bank's funding situation, which is explicitly modelled, plays a crucial role for the specification of bank default. Under defined circumstances, liquidity shortages can be overcome by drawing cash from a collateralized liquidity line. Our model distinguishes between bank and cover pool default and takes into consideration two different default

© Springer Fachmedien Wiesbaden GmbH, part of Springer Nature 2018
M. Spangler, *Modelling German Covered Bonds*, Mathematische Optimierung und Wirtschaftsmathematik | Mathematical Optimization and Economathematics, https://doi.org/10.1007/978-3-658-23915-2_7

triggering events: overindebtedness and illiquidity. While overindebtedness is caused by a deterioration of asset quality, illiquidity stems from the inability to raise sufficient funding to fulfil payment obligations. Both default events are influenced by the market environment and the resulting liquidation payments take into account the Pfandbrief-specific priority of payments.

The model's primary outputs are Pfandbrief default statistics. If adequately calibrated it can therefore be used for the estimation of real-world default probabilities and losses given default, as a supplement to and a verification of existing expert judgement for these parameters. This is especially desirable given the lack of historic default data, but results have to be interpreted with care as this also implies that backtesting is not possible. Nonetheless, our simulation results obtained from an exemplary Mortgage Pfandbrief calibration with typical asset liability mismatches capture the main expected behaviour patterns of Pfandbriefe. Even if there is a non-negligible probability of bank default after the first few years, Pfandbriefe are still comparably safe with much lower default probabilities, at least for short to mid-term maturities. Longer-dated Pfandbriefe are affected by time subordination, which arises from the fact that the PfandBG does not specify any restrictions to use cover pool proceeds for the repayment of Pfandbriefe with shorter maturities. Due to the priority claim on the cover pool, Pfandbrief recovery rates are notably higher than those of other liabilities which suffer from structural subordination and potential asset encumbrance. The value of the Pfandbrief holder's additional claim on the bank's general insolvency estate depends on the amount of unencumbered other assets at the time of bank default. Our sensitivity analyses reveal that the impact of model parameters on resulting default statistics can be quite high, which implies that the individual risk profile plays an important role and that Pfandbriefe are not automatically safe by construction. To obtain realistic bank-specific results, an adequate model calibration is therefore of uttermost importance.

Our Pfandbrief model provides a flexible framework for structural analyses and can be easily extended for tailor-made analyses. It can, for example, contribute to current debates on extendible Pfandbrief maturities. As outlined in vdp [156], there is a proposal to introduce a maturity extension option of up to 12 months for newly issued and outstanding Pfandbriefe, to ease funding pressure and to reduce the risk of fire-sales in times of market distortions in the immediate aftermath of an issuer default. Unlike soft bullet bonds in other jurisdictions which have become more and more popular in recent years, the proposed limited maturity extension is planned to be not trigger-based but needs to be activated by the cover pool administrator. At the time of writing, the final design of the maturity extension was not yet clear and discussions were still ongoing, cf. IFR [95]. Once potential criteria for the maturity extension are specified and implemented in our model, a quantification of the impact on Pfandbrief default statistics is straightforward. The insights of such analyses would be especially interesting in light of market observations which do not reveal a clear spread differentiation between soft bullet and hard bullet structures, cf. Rudolf and Rühlmann [138]. Other interesting topics which may be worth further investigations include but are not limited to (i) a comparison of different business models through balance sheet modifications and variations in asset liability management strategies, (ii) a quantification of the value of individual legal cover requirements, (iii) analyses related to cover pool insolvency timing, (iv) analyses in the context of asset encumbrance caused by Pfandbrief issuance and (v) analyses regarding the cover add-on which has been introduced in 2014 to facilitate a reaction of BaFin to unfavourable issuer- or cover pool-specific developments

We conclude this work by identifying several avenues for potential model improvements. First, our model assumes that it is sufficient to model only one type of Pfandbriefe, the Pfandbrief type under consideration. In practice, Pfandbrief banks often have more than one type of Pfandbriefe outstanding. As long as the size of the additional Pfandbrief business is small, our assumptions in this context (i.e. an implicit modelling by adjusting funding and liquidation haircuts in a suitable manner) may be viable, but for banks with considerable volumes of other Pfandbrief types an explicit modelling of these positions should be considered to allow for a proper study of cross-effects between the different cover pools and their impact on the bank's creditors. Even though the modelling of such effects is likely to have an impact on the resulting default statistics, we expect our main insights to remain unchanged. Second, our modelling approach relies on strong assumptions in the context of default specification. The choice of our bank default barrier which triggers overindebtedness is not straightforward and may have a considerable impact on model results. Given empirical findings and our exemplary model calibration, we could not find any indications that our default barrier choice is inadequate. Nevertheless, further analyses regarding the choice of the default barrier, especially in light of the prevalence of a second default reason (illiquidity), are worth consideration. This also applies to the default barrier of the cover pool, which is chosen in a similar way as in the bank's case, due to the fact that the PfandBG does not specify concrete default triggering events for the cover pool. Third, further analyses on the choice of the asset model would also be desirable. The current model allows for analytical solutions for asset present values under stochastic interest rates and creditworthiness (a necessary requirement), but thorough analyses regarding the advantages and implications of different asset models and the choice of the most suitable model for our purpose was beyond the scope of this work. Given the limited degrees of freedom (only three parameters), our asset model cannot be reasonably calibrated to a term structure of PDs, LGDs and spreads, which describes how these parameters are expected to change over time with decreasing time to maturity. In addition, findings from our analyses indicate that the used asset model is likely to be not suitable for arbitrary constellations of asset default probabilities and losses given default as extreme state variable process parameters obtained in the calibration are likely to result in simulation instabilities. The Vasicek model underlying the real-world and risk-neutral dynamics of the interest rates also reveals some deficiencies in the context of calibration, which allow for a questioning of the related model assumptions. Finally, extending our model to allow for more sophisticated asset liability management would enable a proper study of how assumptions in the context of funding, reinvestments and the maintenance of overcollateralization affect the default statistics. In practice, asset liability management decisions depend not only on the financial situation of the bank and on legal requirements but also on strategic considerations. Taking all relevant aspects into account, the modelling can quickly become very complex, which means that a reasonable trade-off between adequacy and resulting model complexity is necessary. In our model, two assumptions in the context of asset liability management are particularly worth further investigations. We currently assume that voluntary overcollateralization in excess of legal requirements is not actively maintained. Here, it would be interesting to see how a certain level of voluntary overcollateralization impacts the risk profile of Pfandbriefe and other liabilities. This would, however, also need further assumptions regarding the refinancing of the additional overcollateralization, which becomes especially relevant when there is not sufficient cash available to add additional cover pool assets. The second assumption which might be reviewed is related to the bank's and the cover pool's reinvestment strategies. In our model, the bank's reinvestment activities concentrate on other assets and always take place pro rata. Here, more realistic strategies which account for current asset performance and existing asset liability mismatches or a given target asset mix could be considered. The same holds for the cover pool administrator's reinvestment strategy, which currently only concentrates on liquid cover pool assets.

In the end, no model can fully explain the complex interdependencies on financial markets. Nevertheless, a Pfandbrief model like ours can help to analyse and understand certain relationships better.

Bibliography

[1] Aareal Bank AG. Pfandbrief statistics pursuant to §28 Pfandbrief Act, Q4 2014.

[2] V. Acharya, D. Gale, and T. Yorulmazer. Rollover Risk and Market Freezes. NBER Working Paper No. 15674, January 2010.

[3] V. Acharya, J.-Z. Huang, M. Subrahmanyam, and R. K. Sundaram. When does strategic debt-service matter? Economic Theory, Vol. 29, Issue 2, October 2006, pp. 363–378.

[4] T. Ahnert, K. Anand, P. Gai, and J. Chapman. Safe, or not safe? Covered bonds and Bank Fragility. In: Beiträge zur Jahrestagung des Vereins für Socialpolitik 2015: Ökonomische Entwicklung - Theorie und Politik - Session: Finance and Banks II, No. F09-V3.

[5] E. Altman, A. Resti, and A. Sironi. Default Recovery Rates in Credit Risk Modeling: A Review of the Literature and Empirical Evidence. NYU Working Paper No. S-CDM-03-11, December 2003.

[6] R. Anderson and S. Sundaresan. A comparative study of structural models of corporate bond yields: An exploratory investigation. Journal of Banking & Finance, Vol. 24, 2000, pp. 255–269.

[7] R. W. Anderson and N. Cakici. The Value of Deposit Insurance in the Presence of Interest Rate and Credit Risk. Financial Markets, Institutions & Instruments, Vol. 8, No. 5, November 1999, pp. 45–62.

[8] R. W. Anderson and A. Carverhill. Liquidity and Capital Structure. CEPR Discussion Paper No. 6044, January 2007.

[9] R. W. Anderson and S. Sundaresan. Design and Valuation of Debt Contracts. Review of Financial Studies, Vol. 9, No. 1, 1996, pp. 37–68.

[10] R. W. Anderson, S. Sundaresan, and P. Tychon. Strategic analysis of contingent claims. European Economic Review, Vol. 40, Issue 3–5, April 1996, pp. 871–881.

© Springer Fachmedien Wiesbaden GmbH, part of Springer Nature 2018
M. Spangler, *Modelling German Covered Bonds*, Mathematische Optimierung
und Wirtschaftsmathematik I Mathematical Optimization and Economathematics,
https://doi.org/10.1007/978-3-658-23915-2

[11] A. Arvanitis and J. Gregory. *Credit: The Complete Guide to Pricing, Hedging and Risk Management.* Risk Books, London, 2001.

[12] A. Asvanunt, M. Broadie, and S. Sundaresan. Managing Corporate Liquidity: Strategies and Pricing Implications. International Journal of Theoretical and Applied Finance, Vol. 14, No. 3, 2011, pp. 369–406.

[13] Basel Committee on Banking Supervision. Basel III: The Liquidity Coverage Ratio and liquidity risk monitoring tools. BIS Publication, January 2013.

[14] Basel Committee on Banking Supervision. Principles for Sound Liquidity Risk Management and Supervision. BIS Publication, September 2008.

[15] M. Baxter and A. Rennie. *Financial calculus: An introduction to derivative pricing.* Cambride University Press, 2003.

[16] Bayerische Landesbank. Pfandbrief statistics pursuant to §28 Pfandbrief Act, Q4 2014.

[17] Berlin-Hannoversche Hypothekenbank AG. Pfandbrief statistics pursuant to §28 Pfandbrief Act, Q4 2014.

[18] T. R. Bielecki and M. Rutkowski. *Credit Risk: Modeling, Valuation and Hedging.* Springer-Verlag, Berlin Heidelberg, 2004.

[19] F. Black and J. C. Cox. Valuing Corporate Securities: Some Effects of Bond Indenture Provisions. The Journal of Finance, Vol. 31, Issue 2, May 1976, pp. 351–367.

[20] F. Black and M. Scholes. The Pricing of Options and Corporate Liabilities. The Journal of Political Economy, Vol. 81, Issue 3, 1973, pp. 637–654.

[21] C. Bluhm, L. Overbeck, and C. Wagner. *An Introduction to Credit Risk Modeling.* Financial Mathematics Series. Chapman & Hall/CRC, London, 2003.

[22] Bremer Landesbank. Pfandbrief statistics pursuant to §28 Pfandbrief Act, Q4 2014.

[23] D. Brigo, A. Dalessandro, M. Neugebauer, and F. Triki. A Stochastic Processes Toolkit for Risk Management. Working Paper, November 2007.

[24] D. Brigo and F. Mercurio. *Interest Rate Models - Theory and Practice: With Smile, Inflation and Credit.* Springer-Verlag, Berlin Heidelberg, 2nd edition, 2006.

[25] E. Briys and F. de Varenne. Valuing Risky Fixed Rate Debt: An Extension. Journal of Financial and Quantitative Analysis, Vol. 32, No. 2, June 1997, pp. 239–248.

[26] M. Broadie, M. Chernov, and S. Sundaresan. Optimal Debt and Equity Values in the Presence of Chapter 7 and Chapter 11. The Journal of Finance, Vol. 62, No. 3, June 2007, pp. 1341–1377.

[27] A. Charitou, N. Lambertides, and L. Trigeorgis. Bankruptcy prediction and structural credit risk models. In: Advances in Credit Risk Modelling and Corporate Bankruptcy Prediction, Edited by Stewart Jones and David A. Hensher, Cambridge University Press, 2008.

[28] R.-R. Chen, N. K. Chidambaran, M. B. Imerman, and B. J. Sopranzetti. Anatomy of Financial Institutions in Crisis: Endogenous Modeling of Bank Default Risk. Working Paper, January 2012.

[29] R.-R. Chen, S.-Y. Hu, and G.-G. Pan. Default Prediction of Various Structural Models. Working Paper, Rutgers University, National Taiwan University, and National Ping-Tung University of Sciences and Technologies, July 2006.

[30] P. Collin-Dufresne and R. S. Goldstein. Do Credit Spreads Reflect Stationary Leverage Ratios? The Journal of Finance, Vol. 56, No. 5, October 2001, pp. 1929–1957.

[31] P. Collin-Dufresne, R. S. Goldstein, and J. S. Martin. The Determinants of Credit Spread Changes. The Journal of Finance, Vol. 56, No. 6, December 2001, pp. 2177–2207.

[32] Commerzbank AG. Pfandbrief statistics pursuant to §28 Pfandbrief Act, Q4 2014.

[33] Committee on the Global Financial System. Asset encumbrance, financial reform and the demand for collateral assets. CGFS Papers No 49, BIS Publication, May 2013.

[34] Corealcredit Bank AG. Pfandbrief statistics pursuant to §28 Pfandbrief Act, Q4 2014.

[35] P. Crosbie and J. Bohn. Modeling Default Risk. Moody's KMV, Modeling Methodology, December 2003.

[36] S. A. Davydenko. When Do Firms Default? A Study of the Default Boundary. EFA Moscow Meetings Paper, November 2012.

[37] DBRS. Rating European Covered Bonds. Methodology, January 2014.

[38] DekaBank Deutsche Girozentrale. Pfandbrief statistics pursuant to §28 Pfandbrief Act, Q4 2014.

[39] Deutsche Apotheker- und Ärztebank eG. Pfandbrief statistics pursuant to §28 Pfandbrief Act, Q4 2014.

[40] Deutsche Bank. The Pfandbrief and other covered bonds in the EU capital market: development and regulation. Deutsche Bank Research, October 2003.

[41] Deutsche Bundesbank. Ergebnisse der fünften Auswirkungsstudie zu Basel II in Deutschland. June 2006.

[42] Deutsche Genossenschafts-Hypothekenbank AG. Pfandbrief statistics pursuant to §28 Pfandbrief Act, Q4 2014.

[43] Deutsche Hypothekenbank. Pfandbrief statistics pursuant to §28 Pfandbrief Act, Q4 2014.

[44] Deutsche Kreditbank AG. Pfandbrief statistics pursuant to §28 Pfandbrief Act, Q4 2014.

[45] Deutsche Pfandbriefbank AG. Pfandbrief statistics pursuant to §28 Pfandbrief Act, Q4 2014.

[46] Deutsche Postbank AG. Pfandbrief statistics pursuant to §28 Pfandbrief Act, Q4 2014.

[47] Deutscher Bundestag. Entwurf eines Gesetzes zur Umsetzung der Richtlinie 2002/47/EG vom 6. Juni 2002 über Finanzsicherheiten und zur Änderung des Hypothekenbankgesetzes und anderer Gesetze. Drucksache 15/1853, 29.10.2003.

[48] Dexia Kommunalbank Deutschland AG. Pfandbrief statistics pursuant to §28 Pfandbrief Act, Q4 2014.

[49] G. Dionne and S. Laajimi. On the determinants of the implied default barrier. Journal of Empirical Finance, Vol. 19, Issue 3, June 2012, pp. 395–408.

[50] L. Dubrana. A Stochastic Model for Credit Spreads Under a Risk-Neutral Framework Through the Use of an Extended Version of the Jarrow, Lando and Turnbull Model. SSRN Working Paper, June 2011.

[51] D. Duffie and K. J. Singleton. Modeling Term Structures of Defaultable Bonds. The Review of Financial Studies, Vol. 12, No. 4, 1999, pp. 687–720.

[52] D. Duffie and K. J. Singleton. *Credit Risk: Pricing, Measurement, and Management.* Princenton Series in Finance. Princeton University Press, 2003.

[53] Düsseldorfer Hypothekenbank AG. Pfandbrief statistics pursuant to §28 Pfandbrief Act, Q4 2014.

[54] DVB Bank SE. Pfandbrief statistics pursuant to §28 Pfandbrief Act, Q4 2014.

[55] A. Elizalde. Credit Risk Models I: Default Correlation in Intensity Models. CEMFI Working Paper No. 0605, April 2006.

[56] A. Elizalde. Credit Risk Models II: Structural Models. CEMFI Working Paper No. 0606, April 2006.

[57] A. Elizalde. Credit Risk Models III: Reconciliation Reduced – Structural Models. CEMFI Working Paper No. 0607, April 2006.

[58] E. J. Elton, M. J. Gruber, D. Agrawal, and C. Mann. Explaining the Rate Spread on Corporate Bonds. The Journal of Finance, Vol. 56, Issue 1, February 2001, pp. 247–277.

[59] Y. H. Eom, J. Helwege, and J.-Z. Huang. Structural Models of Corporate Bond Pricing: An Empirical Analysis. Review of Financial Studies, Vol. 17, No. 2, 2004, pp. 499–544.

[60] J. Ericsson and J. Reneby. The Valuation of Corporate Liabilities: Theory and Tests. SSE/EFI Working Paper Series in Economics and Finance, No. 445, January 2003.

[61] M. Escobar, T. Friedrich, L. Seco, and R. Zagst. A General Structural Approach For Credit Modeling Under Stochastic Volatility. Capco Journal of Financial Transformation, Vol. 32, August 2011, pp. 123–132.

[62] European Covered Bond Council. European Covered Bond Fact Book. ECBC Publication, 11th edition, August 2016.

[63] European Covered Bond Council. European Covered Bond Fact Book. ECBC Publication, 7th edition, September 2012.

[64] European Covered Bond Council. European Covered Bond Fact Book. ECBC Publication, 8th edition, September 2013.

[65] European Covered Bond Council. European Covered Bond Fact Book. ECBC Publication, 9th edition, September 2014.

[66] European Covered Bond Council. European Covered Bond Fact Book. ECBC Publication, 10th edition, August 2015.

[67] F. J. Fabozzi, editor. *Handbook of Finance: Financial Markets and Instruments*, volume 1. John Wiley & Sons, Inc., Hoboken, New Jersey, 2008.

[68] F. J. Fabozzi, editor. *Handbook of Finance: Valuation, Financial Modeling and Quantitative Tools*, volume 3. John Wiley & Sons, Inc., Hoboken, New Jersey, 2008.

[69] F. J. Fabozzi, R.-R. Chen, S.-Y. Hu, and G.-G. Pan. Tests of the performance of structural models in bankruptcy prediction. The Journal of Credit Risk, Vol. 6, No. 2, Summer 2010, pp. 37–78.

[70] H. Fan and S. Sundaresan. Debt Valuation, Renegotiation, and Optimal Dividend Policy. The Review of Financial Studies, Vol. 13, No. 4, 2000, pp. 1057 – 1099.

[71] E. O. Fischer, R. Heinkel, and J. Zechner. Dynamic Capital Structure Choice: Theory and Tests. The Journal of Finance, Vol. 44, No. 1, March 1989, pp. 19–40.

[72] P. François and E. Morellec. Capital Structure and Asset Prices: Some Effects of Bankruptcy Procedures. Journal of Business, Vol. 77, No. 2, April 2004, pp. 387–411.

[73] J. Frye. Depressing recoveries. Risk, November 2000, pp. 108–111.

[74] D. Galai, A. Raviv, and Z. Wiener. Liquidation triggers and the valuation of equity and debt. Journal of Banking & Finance, Vol. 31, No. 12, December 2007, pp. 3604–3620.

[75] R. Geske. The Valuation Of Compound Options. Journal of Financial Economics, Vol. 7, No. 1, March 1979, pp. 63–81.

[76] R. Geske. The Valuation Of Corporate Liabilities As Compound Options. Journal of Financial and Quantitative Analysis, Vol. 12, No. 4, November 1977, pp. 541–552.

[77] R. Geske and H. E. Johnson. The Valuation of Corporate Liabilities as Compound Options: A Correction. Journal of Financial and Quantitative Analysis, Vol. 19, No. 2, June 1984, pp. 231–232.

[78] P. Glassermann. *Monte Carlo Methods in Financial Engineering*. Springer Verlag, 2003.

[79] R. S. Goldstein, N. Ju, and H. Leland. An EBIT-Based Model of Dynamic Capital Structure. Journal of Business, Vol. 74, No. 4, October 2001, pp. 483–512.

[80] R. Grossmann and O. Stöcker. Overview of covered bonds. In: European Covered Bond Council: *European Covered Bond Fact Book*, ECBC Publication, 11th edition, August 2016, pp. 131–144.

[81] X. Guo, R. A. Jarrow, and Y. Zeng. Credit Risk Models with Incomplete Information. Mathematics of Operations Research, Vol. 34, No. 2, May 2009, pp. 320–332.

[82] L. Hagen and R. Holter. Auswirkungen von Basel II auf das Risikomanagement deutscher Hyopthekenbanken. In: Verband Deutscher Hypothekenbanken, Professionelles Immobilien-Banking, 1st edition, Berlin, 2002, pp. 53–64.

[83] Hamburger Sparkasse AG. Pfandbrief statistics pursuant to §28 Pfandbrief Act, Q4 2014.

[84] F. Hillenbrand. Covered Bonds in der Insolvenz – ein systematischer Überblick. in: Immobilien & Finanzierung 14, 2013.

[85] T. S. Y. Ho and R. F. Singer. Bond indenture provisions and the risk of corporate debt. Journal of Financial Economics, Vol. 10, Issue 4, December 1982, pp. 375–406.

[86] HSH Nordbank AG. Pfandbrief statistics pursuant to §28 Pfandbrief Act, Q4 2014.

[87] J. C. Hsu, J. Saá-Requejo, and P. Santa-Clara. A Structural Model of Default Risk. The Journal of Fixed Income, Vol. 19, No. 3, Winter 2010, pp. 77–94.

[88] J. Huang and M. Huang. How Much of the Corporate-Treasury Yield Spread is Due to Credit Risk? Review of Asset Pricing Studies, Vol. 2, No. 2, 2012, pp. 153–202.

[89] M. Hughes and R. Werner. Choosing Markovian Credit Migration Matrices by Nonlinear Optimization. Risks, Vol. 4, No. 3, August 2016.

[90] M. Hughes and R. Werner. On the Valuation of German Covered Bonds – One Period Approaches. Working Paper, 2014.

[91] J. Hull and A. White. An Analysis of the Bias in Option Pricing Caused by a Stochastic Volatility. Advances in Futures and Options Research, Vol. 3, 1988, pp. 29–61.

[92] J. Hull and A. White. Branching Out. Risk 7, 1994, pp. 34–37.

[93] M. B. Imerman. Structural credit risk models in banking with applications to the financial crisis. Dissertation, 2011.

[94] ING-DiBa AG. Pfandbrief statistics pursuant to §28 Pfandbrief Act, Q4 2014.

[95] International Financing Review. Roundtable Covered Bonds. IFR Report, March 2017.

[96] R. A. Jarrow, D. Lando, and S. M. Turnbull. A Markov Model for the Term Structure of Credit Risk Spreads. The Review of Financial Studies, Vol. 10, No. 2, Summer 1997, pp. 481–523.

[97] R. A. Jarrow and P. Protter. Structural Versus Reduced Form Models: A New Information Based Perspective. Journal of Investment Management, Vol. 2, No. 2, May 2004, pp. 1–10.

[98] R. A. Jarrow and S. M. Turnbull. Pricing Derivatives on Financial Securities Subject to Credit Risk. The Journal of Finance, Vol. 50, No. 1, March 1995, pp. 53–85.

[99] N. Ju, R. Parrino, A. M. Poteshman, and M. S. Weisbach. Horses and Rabbits? Trade-Off Theory and Optimal Capital Structure. Journal of Financial and Quantitative Analysis, Vol. 40, No. 2, June 2005, pp. 259–281.

[100] I. J. Kim, K. Ramaswamy, and S. Sundaresan. Does Default Risk in Coupons Affect the Valuation of Corporate Bonds?: A Contingent Claims Model. Financial Management, Vol. 22, No. 3, Autumn 1993, pp. 117–131.

[101] J. King and F. Will. The repo treatment of covered bonds by central banks. In: European Covered Bond Council: *European Covered Bond Fact Book*, ECBC Publication, 9th edition, September 2014, pp. 132–163.

[102] T. Koppmann. *Gedeckte Schuldverschreibungen in Deutschland und Großbritanien: Pfandbriefe und UK Covered Bonds im Rechtsvergleich.* Schriften zum Europäischen und Internationalen Privat-, Bank- und Wirtschaftsrecht (31). De Gruyter Verlag, Berlin, 2009.

[103] Kreissparkasse Köln. Pfandbrief statistics pursuant to §28 Pfandbrief Act, Q4 2014.

[104] S. Laajimi. Structural Credit Risk Models: A Review. Insurance and Risk Management, Vol. 80, No. 1, April 2012, pp. 53–93.

[105] Landesbank Baden-Württemberg. Pfandbrief statistics pursuant to §28 Pfandbrief Act, Q4 2014.

[106] Landesbank Berlin. Pfandbrief statistics pursuant to §28 Pfandbrief Act, Q4 2014.

[107] Landesbank Hessen-Thüringen. Pfandbrief statistics pursuant to §28 Pfandbrief Act, Q4 2014.

[108] D. Lando. *Credit Risk Modeling: Theory and Applications.* Princenton Series in Finance. Princeton University Press, 2004.

[109] H. Langer and A. Schadow. Pre-insolvency proceedings and regulatory changes – challenging times for covered bonds. In: European Covered Bond Council: *European Covered Bond Fact Book*, ECBC Publication, 8th edition, September 2013, pp. 87–93.

[110] Lebensausschuss der DAV. *Stochastisches Unternehmensmodell für deutsche Lebensversicherungen: Abschlussbericht der DAV-Arbeitsgruppe.* Verlag Versicherungswirtschaft GmbH Karlsruhe, 2005.

[111] H. E. Leland. Corporate Debt Value, Bond Covenants, and Optimal Capital Structure. The Journal of Finance, Vol. 49, No. 4, September 1994, pp. 1213–1252.

[112] H. E. Leland. Predictions of Default Probabilities in Structural Models of Debt. Working Paper, April 2004.

[113] H. E. Leland and K. B. Toft. Optimal Capital Structure, Endogenous Bankruptcy, and the Term Structure of Credit Spreads. The Journal of Finance, Vol. 51, No. 3, July 1996, pp. 987–1019.

[114] G. Liang, E. Lütkebohmert, and Y. Xiao. A Multi-Period Bank Run Model for Liquidity Risk. Review of Finance, Vol. 18, No. 2, April 2014, pp. 803–842.

[115] G. Liang, E. Lütkebohmert, and Y. Xiao. A Multi-Period Bank Run Model for Liquidity Risk – Data Supplement. Review of Finance, Vol. 18, No. 2, April 2014.

[116] F. Longstaff and E. Schwartz. A Simple Approach to Valuing Risky Fixed and Floating Rate Debt. The Journal of Finance, Vol. 50, No. 3, July 1995, pp. 789–819.

[117] S. P. Mason and S. Bhattacharya. Risky debt, jump processes, and safety covenants. Journal of Financial Economics, Vol. 9, Issue 3, September 1981, pp. 281 – 307.

[118] F. J. Massey. The Kolmogorov-Smirnov Test for Goodness of Fit. Journal of the American Statistical Association, Vol. 46, No. 253, 1951, pp. 68–78.

[119] L. Matz. Contingency Planning. In: L. Matz & P. Neu (ed.): *Liquidity Risk: Measurement and Management*, John Wiley & Sons (Asia) Pte Ltd, Singapore, 2007, pp. 121–145.

[120] L. Matz and P. Neu. Liquidity Risk Management Strategies and Tactics. In: L. Matz & P. Neu (ed.): *Liquidity Risk: Measurement and Management*, John Wiley & Sons (Asia) Pte Ltd, Singapore, 2007, pp. 100–120.

[121] L. Matz and P. Neu, editors. *Liquidity Risk: Measurement and Management*. John Wiley & Sons (Asia) Pte Ltd, Singapore, 2007.

[122] P. Mella-Barral. The Dynamics of Default and Debt Reorganization. The Review of Financial Studies, Vol. 12, No. 3, 1999, pp. 535–578.

[123] P. Mella-Barral and W. Perraudin. Strategic Debt Service. The Journal of Finance, Vol. 52, No. 2, June 1997, pp. 531–556.

[124] R. C. Merton. On the Pricing of Corporate Debt: The Risk Structure of Interest Rates. The Journal of Finance, Vol. 29, Issue 2, 1974, pp. 449–470.

[125] M.M.Warburg & CO Hypothekenbank AG. Pfandbrief statistics pursuant to §28 Pfandbrief Act, Q4 2014.

[126] F. Moraux. Valuing Corporate Liabilities When the Default Threshold is not an Absorbing Barrier. Working Paper, University of Rennes, 2002.

[127] E. Morellec. Asset liquidity, capital structure, and secured debt. Journal of Financial Economics, Vol. 61, Issue 2, August 2001, pp. 173–206.

[128] S. Morris and H. S. Shin. Illiquidity Component of Credit Risk. Working Paper, Princeton University, February 2010.

[129] C. Muñoz and B. Mezza. Fitch Ratings' covered bonds rating methodology. In: European Covered Bond Council: *European Covered Bond Fact Book*, ECBC Publication, 8th edition, September 2013, pp. 515–522.

[130] Münchener Hypothekenbank eG. Geschäftsbericht 2014. 2015.

[131] Münchener Hypothekenbank eG. Pfandbrief statistics pursuant to §28 Pfandbrief Act, Q4 2014.

[132] Natixis Pfandbriefbank AG. Pfandbrief statistics pursuant to §28 Pfandbrief Act, Q4 2014.

[133] Nord/LB Norddeutsche Landesbank Girozentrale. Pfandbrief statistics pursuant to §28 Pfandbrief Act, Q4 2014.

[134] A. Paseka. Debt Valuation with Endogenous Default and Chapter 11 Reorganization. Working Paper, University of Arizona, June 2003.

[135] W. R. M. Perraudin and A. P. Taylor. Liquidity and Bond Market Spreads. EFA 2003 Annual Conference Paper No. 879, June 2003.

[136] M. Pollmann. Das neue Pfandbriefrecht in Deutschland. Diplomarbeit, Fachhochschule der Deutschen Bundesbank, Diplomica Verlag, Hamburg, 2006.

[137] M. Prokopczuk and V. Vonhoff. Risk Premia in Covered Bond Markets. Journal of Fixed Income, Vol. 22, No. 2, pp. 19–29, 2012.

[138] F. Rudolf and K. Rühlmann. Extendable maturity structures - the new normal? In: European Covered Bond Council: *European Covered Bond Fact Book*, ECBC Publication, 11th edition, August 2016, pp. 80–90.

[139] M. Rudolf and A. Saunders. Refinancing Real Estate Loans – Lessons to be Learned from the Subprime Crisis. vdp series of papers, Berlin, 2009.

[140] Saar LB. Pfandbrief statistics pursuant to §28 Pfandbrief Act, Q4 2014.

[141] P. J. Schönbucher. Valuation of Securities Subject to Credit Risk. Bonn University, Working Paper, 1996.

[142] P. J. Schönbucher. *Credit Derivatives Pricing Models: Models, Pricing and Implementation*. Wiley Finance Series. John Wiley & Sons, 2003.

[143] SEB AG. Pfandbrief statistics pursuant to §28 Pfandbrief Act, Q4 2014.

[144] J. B. Siewert and V. Vonhoff. Liquidity and Credit Risk Premia in the Pfandbrief Market. SSRN Working Paper, September 2011.

[145] M. Spangler and R. Werner. *German Covered Bonds: Overview and Risk Analysis of Pfandbriefe*. SpringerBriefs in Finance, 2014.

[146] Sparkasse KölnBonn. Pfandbrief statistics pursuant to §28 Pfandbrief Act, Q4 2014.

[147] R. K. Sundaram. The Merton/KMV Approach to Pricing Credit Risk. Working Paper, January 2001.

[148] R. Sünderhauf. *Bewertung des Ausfallrisikos deutscher Hypothekenbank-Pfandbriefe*. BWV Berliner Wissenschafts-Verlag GmbH, 2006.

[149] D. Tasche. Fitting a distribution to Value-at-Risk and Expected Shortfall, with an application to covered bonds. Journal of Credit Risk, Vol. 12, No. 2, pp. 1–34, 2016.

[150] M. Uhrig-Homburg. Cash-flow shortage as an endogenous bankruptcy reason. Journal of Banking & Finance, Vol. 29, Issue 6, June 2005, pp. 1509–1534.

[151] M. Uhrig-Homburg. Valuation of Defaultable Claims - A Survey. Schmalenbach Business Review, Vol. 54, January 2002, pp. 24–57.

[152] UniCredit Bank AG. Pfandbrief statistics pursuant to §28 Pfandbrief Act, Q4 2014.

[153] Valovis Bank AG. Pfandbrief statistics pursuant to §28 Pfandbrief Act, Q4 2014.

[154] O. Vasicek. An Equilibrium Characterization Of The Term Structure. Journal of Financial Economics, Vol. 5, No. 2, November 1977, pp. 177–188.

[155] O. A. Vasicek. Credit Valuation. KMV Publication, March 1984.

[156] Verband Deutscher Pfandbriefbanken. Infobrief, Ausgabe Q2 2016. Available on http://www.pfandbrief.de/.

[157] Verband Deutscher Pfandbriefbanken. Pfandbrief Act (as at November 2015). Brochure.

[158] Verband Deutscher Pfandbriefbanken. The Pfandbrief 2012/2013: Facts and Figures about Europe's Covered Bond Benchmark. Pfandbrief Fact Book, 17th edition, Berlin, 2012.

[159] Verband Deutscher Pfandbriefbanken. The Pfandbrief 2013/2014: Facts and Figures about Europe's Covered Bond Benchmark. Pfandbrief Fact Book, 18th edition, Berlin, 2013.

[160] Verband Deutscher Pfandbriefbanken. The Pfandbrief 2014/2015: Facts and Figures about Europe's Covered Bond Benchmark. Pfandbrief Fact Book, 19th edition, Berlin, 2014.

[161] Verband Deutscher Pfandbriefbanken. The Pfandbrief 2015/2016: Facts and Figures about Europe's Covered Bond Benchmark. Pfandbrief Fact Book, 20th edition, Berlin, 2015.

[162] Verband Deutscher Pfandbriefbanken. The Pfandbrief 2016/2017: Facts and Figures about Europe's Covered Bond Benchmark. Pfandbrief Fact Book, 21th edition, Berlin, 2016.

[163] Verband Deutscher Pfandbriefbanken. Volume outstanding and issuance 2003-2014. Pfandbrief market statistics, 2015.

[164] Verband deutscher Pfandbriefbanken e. V. Pfandbrief statistics pursuant to §28 Pfandbrief Act, Q4 2014.

[165] Westdeutsche ImmobilienBank AG. Pfandbrief statistics pursuant to §28 Pfandbrief Act, Q4 2014.

[166] WL BANK AG Westfälische Landschaft Bodenkreditbank. Pfandbrief statistics pursuant to §28 Pfandbrief Act, Q4 2014.

[167] Wüstenrot Bank AG Pfandbriefbank. Pfandbrief statistics pursuant to §28 Pfandbrief Act, Q4 2014.

[168] R. Zagst. *Interest-Rate Management.* Springer Verlag Berlin Heidelberg, 2002.

[169] C. Zhou. The term structure of credit spreads with jump risk. Journal of Banking & Finance, Vol. 25, No. 11, November 2001, pp. 2015–2040.

Supplementary Tables and Figures

Table A.1: Exemplary bank balance sheet.

t	CPS	CPL	OA	PB	OL
0.5	879	821	5,142	1,700	4,909
1	1,372	—	496	1,054	2,187
1.5	858	—	142	2,155	778
2	1,238	—	142	430	778
2.5	1,082	—	142	309	778
3	1,101	—	142	313	778
3.5	955	—	142	343	778
4	971	—	142	347	778
4.5	872	—	142	572	778
5	887	—	142	579	778
5.5	772	—	919	500	525
6	785	—	919	506	525
6.5	798	—	919	513	525
	810	—	919	520	525
7.5	823	—	919	527	525
8	836	—	919	534	525
8.5	849	—	919	541	525
9	863	—	919	548	525
9.5	876	—	919	555	525
10	890	—	919	563	525
10.5	910	—	919	1,573	525
11	924	—	919	1,594	525
11.5	939	—	919	1,614	525
12	953	—	919	1,636	525
12.5	968	—	919	1,657	525

© Springer Fachmedien Wiesbaden GmbH, part of Springer Nature 2018
M. Spangler, *Modelling German Covered Bonds*, Mathematische Optimierung
und Wirtschaftsmathematik | Mathematical Optimization and Economathematics,
https://doi.org/10.1007/978-3-658-23915-2

Table A.2: Exemplary lifetime default probabilities $\pi_R(t)$ for different rating classes $R \in \{AAA, AA, AA, A+, ..., CCC, D\}$ and asset maturities $t \in \{1, 2, 3, ..., 13\}$.

t	π_{AAA}	π_{AA}	π_{A+}	π_A	π_{A-}	π_{BBB+}	π_{BBB}	π_{BBB-}
1	0.01%	0.02%	0.03%	0.05%	0.09%	0.12%	0.17%	0.26%
2	0.02%	0.04%	0.06%	0.11%	0.19%	0.26%	0.38%	0.58%
3	0.04%	0.07%	0.10%	0.17%	0.31%	0.44%	0.63%	0.98%
4	0.06%	0.10%	0.14%	0.25%	0.45%	0.65%	0.94%	1.45%
5	0.08%	0.14%	0.19%	0.33%	0.61%	0.89%	1.31%	2.01%
6	0.10%	0.18%	0.24%	0.44%	0.81%	1.19%	1.74%	2.66%
7	0.13%	0.22%	0.30%	0.55%	1.04%	1.53%	2.24%	3.38%
8	0.16%	0.28%	0.37%	0.69%	1.30%	1.92%	2.79%	4.18%
9	0.20%	0.34%	0.46%	0.85%	1.60%	2.35%	3.41%	5.05%
10	0.24%	0.41%	0.55%	1.04%	1.94%	2.84%	4.08%	5.97%
11	0.29%	0.50%	0.66%	1.24%	2.31%	3.37%	4.80%	6.94%
12	0.35%	0.59%	0.79%	1.48%	2.73%	3.94%	5.57%	7.95%
13	0.42%	0.70%	0.93%	1.74%	3.18%	4.55%	6.37%	8.99%

t	π_{BB+}	π_{BB}	π_{BB-}	π_{B+}	π_B	π_{B-}	π_{CCC}	π_D
1	0.59%	0.88%	1.98%	2.96%	6.67%	10.00%	20.00%	100%
2	1.33%	1.98%	4.37%	6.52%	13.69%	18.78%	32.76%	100%
3	2.23%	3.30%	7.05%	10.32%	20.23%	26.22%	41.51%	100%
4	3.28%	4.80%	9.89%	14.15%	26.05%	32.45%	47.88%	100%
5	4.47%	6.44%	12.79%	17.88%	31.13%	37.68%	52.75%	100%
6	5.78%	8.19%	15.67%	21.44%	35.57%	42.11%	56.63%	100%
7	7.18%	10.00%	18.50%	24.79%	39.45%	45.88%	59.81%	100%
8	8.64%	11.85%	21.22%	27.91%	42.85%	49.14%	62.47%	100%
9	10.15%	13.71%	23.82%	30.82%	45.86%	51.98%	64.75%	100%
10	11.69%	15.56%	26.29%	33.51%	48.54%	54.47%	66.73%	100%
11	13.24%	17.39%	28.64%	36.01%	50.94%	56.69%	68.46%	100%
12	14.79%	19.19%	30.85%	38.33%	53.09%	58.67%	70.00%	100%
13	16.33%	20.94%	32.94%	40.48%	55.05%	60.46%	71.37%	100%

Table A.3: Calibrated model parameters $\hat{Z}_i^{x,0}$, $\hat{\mu}_i^x$ and $\hat{\sigma}_i^x$ for $i = 1, ..., n_x$ and $x \in \{CPS, OA\}$, based on the Vasicek parameters in Table 5.9 and the correlations in Table 5.5.

t	$\hat{Z}_i^{CPS,0}$	$\hat{\mu}_i^{CPS}$	$\hat{\sigma}_i^{CPS}$	$\hat{Z}_i^{OA,0}$	$\hat{\mu}_i^{OA}$	$\hat{\sigma}_i^{OA}$
0.5	2.0411	0.0030	0.3514	3718.0747	0.0028	3.0381
1	1.8418	0.0036	0.2332	1189.3540	0.0036	2.0107
1.5	1.7129	0.0041	0.1812	538.0490	0.0041	1.5589
2	1.6300	0.0045	0.1513	316.8214	0.0047	1.3000
2.5	1.5586	0.0049	0.1308	197.5797	0.0051	1.1213
3	1.5042	0.0052	0.1160	137.2294	0.0055	0.9935
3.5	1.4542	0.0055	0.1044	97.4509	0.0059	0.8929
4	1.4132	0.0058	0.0953	73.5089	0.0063	0.8140
4.5	1.3748	0.0061	0.0877	56.2259	0.0066	0.7478
5	1.3419	0.0063	0.0814	44.6908	0.0069	0.6932
5.5	1.3108	0.0065	0.0759	35.9184	0.0072	0.6459
6	1.2834	0.0067	0.0713	29.6324	0.0075	0.6055
6.5	1.2575	0.0069	0.0671	24.6829	0.0078	0.5697
7	1.2341	0.0071	0.0635	20.9447	0.0081	0.5384
7.5	1.2120	0.0072	0.0603	17.9251	0.0083	0.5103
8	1.1917	0.0074	0.0574	15.5497	0.0086	0.4852
8.5	1.1726	0.0075	0.0547	13.5920	0.0089	0.4625
9	1.1549	0.0076	0.0524	12.0014	0.0091	0.4420
9.5	1.1381	0.0077	0.0502	10.6685	0.0094	0.4232
10	1.1224	0.0078	0.0482	9.5568	0.0096	0.4061
10.5	1.1076	0.0078	0.0464	8.6122	0.0099	0.3903
11	1.0936	0.0079	0.0447	7.8070	0.0101	0.3758
11.5	1.0804	0.0079	0.0431	7.1145	0.0104	0.3623
12	1.0679	0.0080	0.0417	6.5136	0.0106	0.3498
12.5	1.0560	0.0080	0.0403	5.9912	0.0108	0.3382

Table A.4: Base run default statistics.

t	π_t^B	$\pi_{t,cum}^B$	L_t^{OL}	EL_t^{OL}	$EL_{t,cum}^{OL}$	π_t^P	$\pi_{t,cum}^P$	L_t^{PB}	EL_t^{PB}	$EL_{t,cum}^{PB}$
0.5	0.0%	0.0%	0.0%	0.0%	0.0%	0.0%	0.0%	0.0%	0.0%	0.0%
1	0.0%	0.0%	68.4%	0.0%	0.0%	0.0%	0.0%	0.0%	0.0%	0.0%
1.5	0.0%	0.0%	0.0%	0.0%	0.0%	0.0%	0.0%	0.0%	0.0%	0.0%
2	0.0%	0.0%	95.2%	0.0%	0.0%	0.0%	0.0%	0.0%	0.0%	0.0%
2.5	0.0%	0.0%	94.3%	0.0%	0.0%	0.0%	0.0%	0.0%	0.0%	0.0%
3	0.0%	0.0%	94.3%	0.0%	0.0%	0.0%	0.0%	0.0%	0.0%	0.0%
3.5	0.0%	0.0%	94.3%	0.0%	0.0%	0.0%	0.0%	0.0%	0.0%	0.0%
4	0.1%	0.1%	94.2%	0.1%	0.1%	0.0%	0.0%	0.0%	0.0%	0.0%
4.5	0.2%	0.3%	90.9%	0.2%	0.3%	0.0%	0.0%	0.0%	0.0%	0.0%
5	0.5%	0.7%	90.5%	0.4%	0.7%	0.0%	0.0%	0.0%	0.0%	0.0%
5.5	0.4%	1.2%	92.0%	0.4%	1.1%	0.0%	0.0%	0.0%	0.0%	0.0%
6	0.5%	1.6%	92.1%	0.4%	1.5%	0.0%	0.0%	0.0%	0.0%	0.0%
6.5	0.5%	2.1%	91.6%	0.5%	1.9%	0.0%	0.0%	0.0%	0.0%	0.0%
7	0.6%	2.7%	91.2%	0.5%	2.4%	0.0%	0.0%	0.0%	0.0%	0.0%
7.5	0.6%	3.2%	91.2%	0.5%	3.0%	0.0%	0.0%	0.0%	0.0%	0.0%
8	0.5%	3.8%	90.7%	0.5%	3.4%	0.0%	0.0%	0.0%	0.0%	0.0%
8.5	0.5%	4.3%	90.6%	0.5%	3.9%	0.0%	0.0%	0.0%	0.0%	0.0%
9	0.6%	4.9%	89.8%	0.5%	4.5%	0.0%	0.0%	50.5%	0.0%	0.0%
9.5	0.6%	5.5%	88.7%	0.5%	5.0%	0.0%	0.0%	45.7%	0.0%	0.0%
10	0.6%	6.1%	87.2%	0.5%	5.5%	0.1%	0.1%	38.1%	0.0%	0.0%
10.5	0.8%	6.9%	86.0%	0.7%	6.2%	0.8%	0.9%	28.1%	0.2%	0.2%
11	1.0%	7.9%	79.0%	0.8%	7.0%	1.9%	2.8%	24.4%	0.5%	0.7%
11.5	1.3%	9.2%	67.2%	0.9%	7.9%	2.5%	5.3%	18.8%	0.5%	1.2%
12	1.0%	10.2%	49.3%	0.5%	8.4%	2.1%	7.4%	11.9%	0.2%	1.4%
12.5	0.2%	10.4%	15.8%	0.0%	8.4%	0.2%	7.6%	1.5%	0.0%	1.4%
min	0.0%	0.0%	0.0%	0.0%	0.0%	0.0%	0.0%	0.0%	0.0%	0.0%
max	1.3%	10.4%	95.2%	0.9%	8.4%	2.5%	7.6%	50.5%	0.5%	1.4%

Table A.5: Base run default timing: number of default events by simulation time ('# D' = total number of bank defaults, '# S' = number of simultaneous cover pool defaults, '# L' = number of later cover pool defaults, '# N' = number of no cover pool defaults in case of bank default).

t	# D	# S	# L	# N
0.5	0	0	0	0
1	15	0	0	15
1.5	0	0	0	0
2	0	0	0	0
2.5	6	0	2	4
3	4	0	1	3
3.5	20	0	5	15
4	61	0	38	23
4.5	170	0	129	41
5	451	0	338	113
5.5	444	0	338	106
6	451	0	327	124
6.5	505	0	378	127
7	550	0	405	145
7.5	554	0	392	162
8	538	0	383	155
8.5	538	0	386	152
9	594	0	454	140
9.5	602	0	443	159
10	602	0	452	150
10.5	788	259	365	164
11	1008	411	377	220
11.5	1316	638	325	353
12	1006	594	22	390
12.5	225	152	0	73
Total	10448	2054	5560	2834
%	100%	20%	53%	27%

Table A.6: Base run PD-ratings.

t	OL	PB
0.5	AAA	AAA
1	AA	AAA
1.5	AAA	AAA
2	AAA	AAA
2.5	AAA	AAA
3	AAA	AAA
3.5	AAA	AAA
4	A+	AAA
4.5	A	AAA
5	BBB+	AAA
5.5	BBB	AAA
6	BBB	AAA
6.5	BBB-	AAA
7	BBB-	AAA
7.5	BBB-	AAA
8	BBB-	AAA
8.5	BBB-	AAA
9	BBB-	AAA
9.5	BBB-	AAA
10	BB+	AAA
10.5	BB+	A
11	BB+	BBB+
11.5	BB+	BBB-
12	BB+	BBB-
12.5	BB+	BBB-

Table A.7: Base run default reasons over time ('# D-B' = total number of bank defaults, '# O-B' = number of bank defaults triggered by overindebtedness, '# I-B' = number of bank defaults triggered by illiquidity, '# D-P' = total number of cover pool defaults, '# O-P' = number of cover pool defaults triggered by overindebtedness, '# I-P' = number of cover pool defaults triggered by illiquidity).

t	# D-B	# O-B	# I-B	# D-P	# O-P	# I-P
0.5	0	0	0	0	0	0
1	15	0	15	0	0	0
1.5	0	0	0	0	0	0
2	0	0	0	0	0	0
2.5	6	0	6	0	0	0
3	4	0	4	0	0	0
3.5	20	0	20	0	0	0
4	61	7	54	0	0	0
4.5	170	5	165	0	0	0
5	451	1	450	0	0	0
5.5	444	1	443	0	0	0
6	451	2	449	0	0	0
6.5	505	1	504	0	0	0
7	550	1	549	0	0	0
7.5	554	0	554	0	0	0
8	538	0	538	0	0	0
8.5	538	0	538	0	0	0
9	594	3	591	2	2	0
9.5	602	26	576	10	10	0
10	602	88	514	71	71	0
10.5	788	442	346	771	530	241
11	1008	911	97	1942	1537	405
11.5	1316	1313	3	2482	2233	249
12	1006	1006	0	2091	2087	4
12.5	225	225	0	245	245	0
Total	10448	4032	6416	7614	6715	899
%	100%	39%	61%	100%	88%	12%

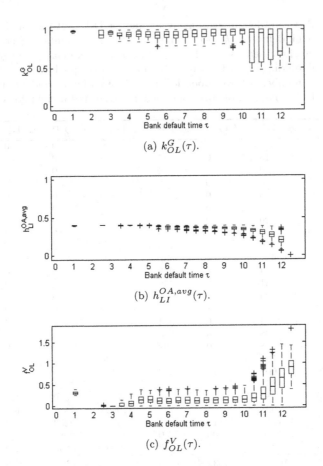

Figure A.1: (part 1) Base run distribution of loss drivers of other liabilities (central mark of boxes: median, edges of boxes: 25th and 75th percentiles).

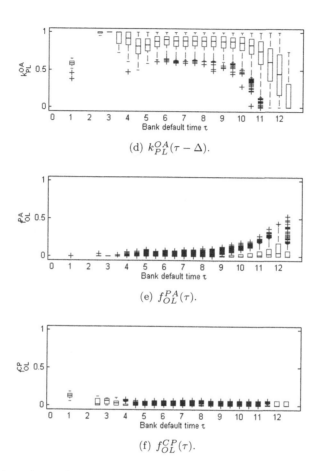

(d) $k_{PL}^{OA}(\tau - \Delta)$.

(e) $f_{OL}^{PA}(\tau)$.

(f) $f_{OL}^{CP}(\tau)$.

Figure A.1: (part 2) Base run distribution of loss drivers of other liabilities (central mark of boxes: median, edges of boxes: 25th and 75th percentiles).

Table A.8: Base run scenarios with the ten highest/lowest losses of other liabilities.

(a) The ten highest losses of other liabilities.

No.	$LS^{OL}(\tau)$	τ	$\frac{V_B^O(\tau)}{B_R(\tau)}$	τ^*	$k_{OL}^G(\tau)$	$h_{LI}^{OA,avg}(\tau)$	$k_{PL}^{OA}(\tau-\Delta)$	$f_{OL}^V(\tau)$	$f_{OL}^{PA}(\tau)$	$f_{OL}^{CP}(\tau)$
1	100%	4	<1	11	86%	n/a	100%	0%	0%	0%
2	100%	2.5	>1	12	100%	n/a	100%	0%	0%	0%
3	100%	2.5	>1	11.5	95%	n/a	100%	0%	0%	0%
4	100%	4	<1	10	89%	n/a	100%	0%	0%	0%
5	100%	4	>1	11	96%	n/a	100%	0%	0%	0%
6	100%	4	>1	11.5	99%	n/a	100%	0%	0%	0%
7	100%	4	>1	10.5	91%	n/a	100%	0%	0%	0%
8	100%	3	>1	11.5	99%	n/a	100%	0%	0%	0%
9	100%	4	>1	11	95%	n/a	100%	0%	0%	0%
10	100%	4	<1	10.5	88%	n/a	100%	0%	0%	0%
min	100%	2.5	—	10	86%	n/a	100%	0%	0%	0%
max	100%	4.0	—	12	100%	n/a	100%	0%	0%	0%

(b) The ten lowest losses of other liabilities.

No.	$LS^{OL}(\tau)$	τ	$\frac{V_B^O(\tau)}{B_R(\tau)}$	τ^*	$k_{OL}^G(\tau)$	$h_{LI}^{OA,avg}(\tau)$	$k_{PL}^{OA}(\tau-\Delta)$	$f_{OL}^V(\tau)$	$f_{OL}^{PA}(\tau)$	$f_{OL}^{CP}(\tau)$
1	0.2%	12.5	<1	∞	100%	0%	25%	84%	10%	6%
2	0.2%	12.5	<1	12.5	91%	0%	38%	90%	18%	0%
3	0.2%	12.5	<1	12.5	95%	0%	5%	103%	2%	0%
4	0.3%	12.5	<1	12.5	74%	0%	0%	135%	0%	0%
5	0.5%	12.5	<1	∞	100%	0%	6%	96%	2%	1%
6	0.6%	12.5	<1	12.5	88%	0%	9%	109%	3%	0%
7	0.7%	12.5	<1	∞	100%	0%	72%	46%	48%	4%
8	0.9%	12.5	<1	∞	100%	0%	13%	88%	5%	6%
9	1.1%	12.5	<1	12.5	80%	0%	0%	124%	0%	0%
10	1.1%	12.5	<1	∞	100%	0%	0%	93%	0%	6%
min	0.2%	12.5	—	12.5	74%	0%	0%	46%	0%	0%
max	1.1%	12.5	—	∞	100%	0%	72%	135%	48%	6%

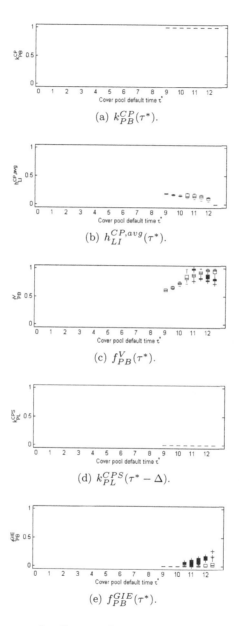

(a) $k_{PB}^{CP}(\tau^*)$.

(b) $h_{LI}^{CP,avg}(\tau^*)$.

(c) $f_{PB}^{V}(\tau^*)$.

(d) $k_{PL}^{CPS}(\tau^* - \Delta)$.

(e) $f_{PB}^{GIE}(\tau^*)$.

Figure A.2: Base run distribution of Pfandbrief loss drivers (central mark of boxes: median, edges of boxes: 25th and 75th percentiles).

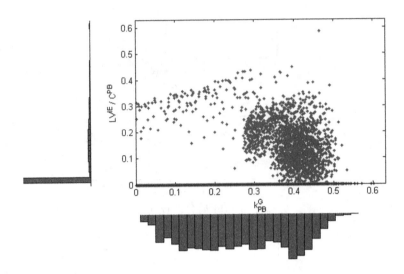

Figure A.3: Bivariate distribution of $k_{PB}^G(\tau^*)$ and $\frac{LV^{IE}(\tau^*)}{C^{PB}(\tau^*)}$ in the base run.

Table A.9: Base run scenarios with the ten highest/lowest losses of Pfandbriefe.

(a) The ten highest Pfandbrief losses.

No.	$LS^{PB}(\tau^*)$	τ^*	$\frac{V_B^O(\tau^*)}{BD(\tau^*)}$	τ	$h_{LI}^{CP,avg}(\tau^*)$	$k_{PL}^{CPS}(\tau^*-\Delta)$	$f_{PB}^V(\tau^*)$	$f_{PB}^{GIE}(\tau^*)$
1	52.8%	9	<1	5.5	20%	0%	59%	0%
2	48.2%	9.5	<1	5	18%	0%	63%	0%
3	48.2%	9.5	<1	5	18%	0%	63%	0%
4	48.1%	9	<1	5	19%	0%	64%	0%
5	47.2%	9.5	<1	5	17%	0%	64%	0%
6	46.3%	9.5	<1	5	16%	0%	64%	0%
7	46.3%	9.5	<1	4.5	17%	0%	65%	0%
8	45.6%	9.5	<1	4	17%	0%	66%	0%
9	45.1%	9.5	<1	5	18%	0%	67%	0%
10	44.0%	9.5	<1	4.5	18%	0%	68%	0%
min	44.0%	9	—	4	16%	0%	59%	0%
max	52.8%	9.5	—	5.5	20%	0%	68%	0%

(b) The ten lowest Pfandbrief losses.

No.	$LS^{PB}(\tau^*)$	τ^*	$\frac{V_B^O(\tau^*)}{BD(\tau^*)}$	τ	$h_{LI}^{CP,avg}(\tau^*)$	$k_{PL}^{CPS}(\tau^*-\Delta)$	$f_{PB}^V(\tau^*)$	$f_{PB}^{GIE}(\tau^*)$
1	0.001%	12.5	<1	12.5	0%	0%	99.9%	0%
2	0.004%	12.5	<1	12.5	0%	0%	98.4%	2%
3	0.005%	12.5	<1	12.5	0%	0%	96.9%	3%
4	0.007%	12.5	<1	12.5	0%	0%	99.7%	0%
5	0.015%	12.5	<1	12.5	0%	0%	99.9%	0%
6	0.016%	12.5	<1	12.5	0%	0%	99.7%	0%
7	0.024%	12.5	<1	12.5	0%	0%	95.7%	4%
8	0.032%	12.5	<1	12.5	0%	0%	99.5%	0%
9	0.033%	12.5	<1	12.5	0%	0%	99.7%	0%
10	0.034%	12.5	<1	12.5	0%	0%	99.3%	1%
min	0.001%	12.5	—	12.5	0%	0%	95.7%	0%
max	0.034%	12.5	—	12.5	0%	0%	99.9%	4%

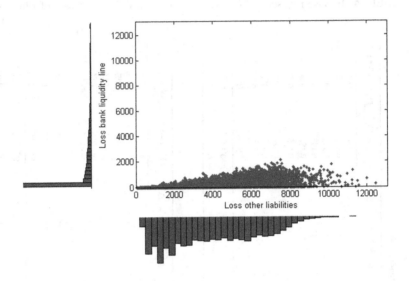

Figure A.4: Bivariate loss distribution of other liabilities and the bank's liquidity line in the base run.

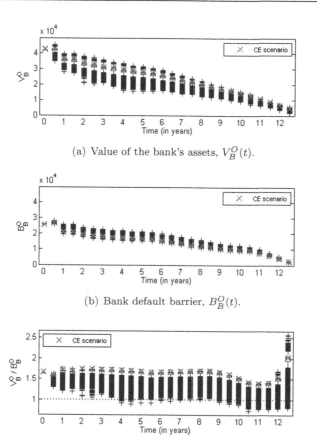

(a) Value of the bank's assets, $V_B^O(t)$.

(b) Bank default barrier, $B_B^O(t)$.

(c) Value of the bank's assets vs. bank default barrier, $\frac{V_B^O(t)}{B_B^O(t)}$.

Figure A.5: The bank's solvency situation in the base run, given that the bank has not defaulted prior to the respective time (central mark of boxes: median, edges of boxes: 25th and 75th percentiles).

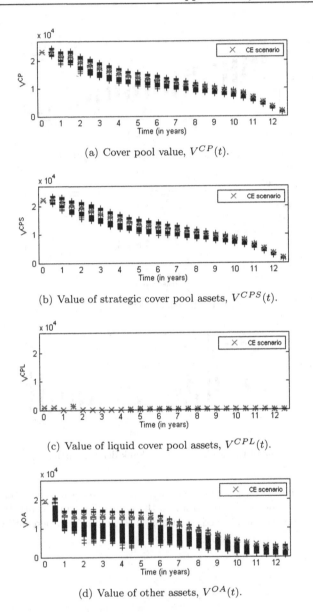

(a) Cover pool value, $V^{CP}(t)$.

(b) Value of strategic cover pool assets, $V^{CPS}(t)$.

(c) Value of liquid cover pool assets, $V^{CPL}(t)$.

(d) Value of other assets, $V^{OA}(t)$.

Figure A.6: Distribution of asset present values in the base run, given that the bank has not defaulted prior to the respective time (central mark of boxes: median, edges of boxes: 25th and 75th percentiles).

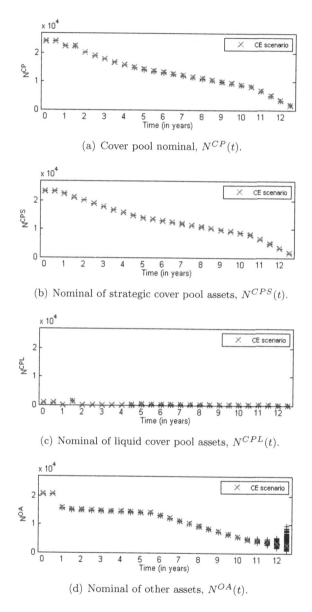

(a) Cover pool nominal, $N^{CP}(t)$.

(b) Nominal of strategic cover pool assets, $N^{CPS}(t)$.

(c) Nominal of liquid cover pool assets, $N^{CPL}(t)$.

(d) Nominal of other assets, $N^{OA}(t)$.

Figure A.7: Distribution of asset nominals in the base run, given that the bank has not defaulted prior to the respective time (central mark of boxes: median, edges of boxes: 25th and 75th percentiles).

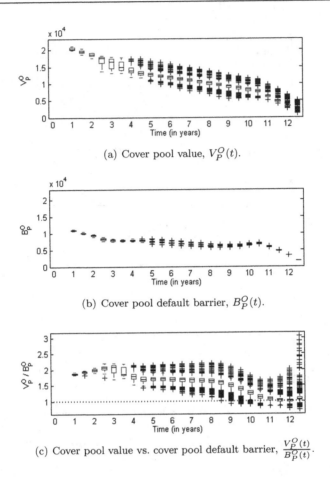

(a) Cover pool value, $V_P^O(t)$.

(b) Cover pool default barrier, $B_P^O(t)$.

(c) Cover pool value vs. cover pool default barrier, $\dfrac{V_P^O(t)}{B_P^O(t)}$.

Figure A.8: The cover pool's solvency situation in the base run, given that the bank has already defaulted but the cover pool has not defaulted prior to the respective time (central mark of boxes: median, edges of boxes: 25th and 75th percentiles).

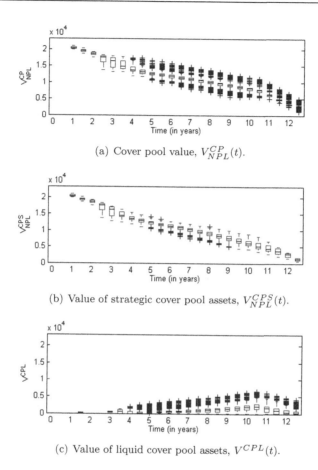

(a) Cover pool value, $V^{CP}_{NPL}(t)$.

(b) Value of strategic cover pool assets, $V^{CPS}_{NPL}(t)$.

(c) Value of liquid cover pool assets, $V^{CPL}(t)$.

Figure A.9: Distribution of cover pool present values in the base run, given that the bank has already defaulted but the cover pool has not defaulted prior to the respective time (central mark of boxes: median, edges of boxes: 25th and 75th percentiles).

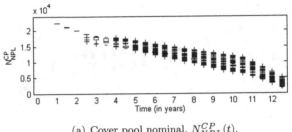

(a) Cover pool nominal, $N_{NPL}^{CP}(t)$.

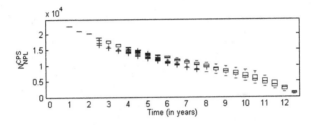

(b) Nominal of strategic cover pool assets, $N_{NPL}^{CPS}(t)$.

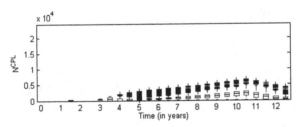

(c) Nominal of liquid cover pool assets, $N^{CPL}(t)$.

Figure A.10: Distribution of cover pool nominals in the base run, given that the bank has already defaulted but the cover pool has not defaulted prior to the respective time (central mark of boxes: median, edges of boxes: 25th and 75th percentiles).

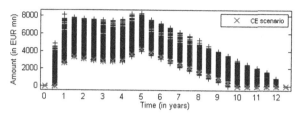

(a) The bank's funding need, $G_B^C(t)$.

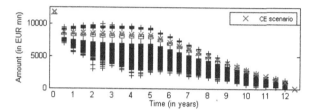

(b) Potential funding from other assets, $L_{OA}^{\max}(t)$.

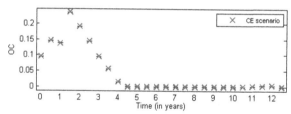

(c) Percentage nominal overcollateralization, $\dfrac{\tilde{N}^{CPS}(t) - \tilde{N}^{PB}(t)}{\tilde{N}^{PB}(t)}$.

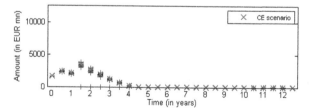

(d) Potential funding from strategic cover pool assets, $L_{CPS}^{\max}(t)$.

Figure A.11: The bank's funding situation in the base run, given that the bank has not defaulted prior to the respective time (central mark of boxes: median, edges of boxes: 25th and 75th percentiles).

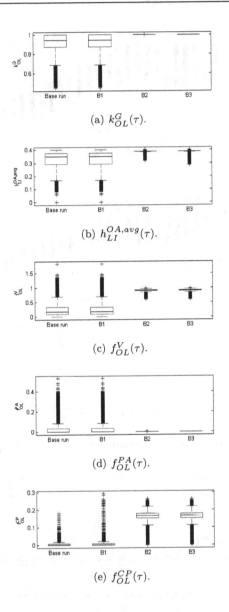

(a) $k_{OL}^G(\tau)$.

(b) $h_{LI}^{OA,avg}(\tau)$.

(c) $f_{OL}^V(\tau)$.

(d) $f_{OL}^{PA}(\tau)$.

(e) $f_{OL}^{CP}(\tau)$.

Figure A.12: Bank funding scenario analysis: distribution of loss drivers of other liabilities (central mark of boxes: median, edges of boxes: 25th and 75th percentiles).

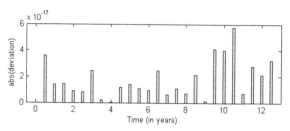

(a) Short rate mean $\mu_{r(t)|\mathcal{F}_0}$: model vs. realized.

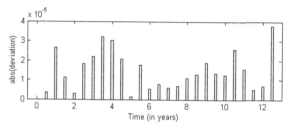

(b) Short rate volatility $\sigma_{r(t)|\mathcal{F}_0}$: model vs. realized.

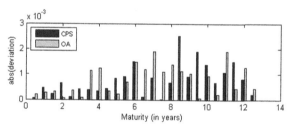

(c) Asset PDs π_i^{CPS} and π_j^{OA}: model vs. realized.

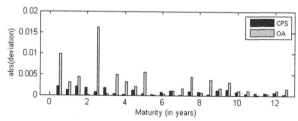

(d) Asset LGDs L_i^{CPS} and L_j^{OA}: model vs. realized.

Figure A.13: Absolute deviations of realized statistics from their theoretical counterparts in the base run, by time bucket.

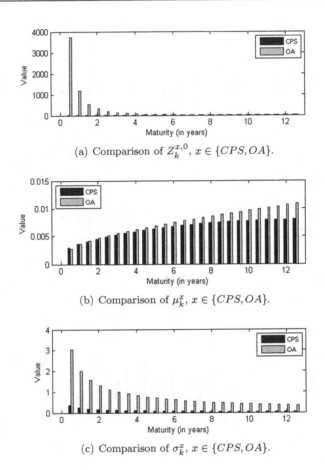

(a) Comparison of $Z_k^{x,0}$, $x \in \{CPS, OA\}$.

(b) Comparison of μ_k^x, $x \in \{CPS, OA\}$.

(c) Comparison of σ_k^x, $x \in \{CPS, OA\}$.

Figure A.14: State variable process parameters in the base run.

Printed in the United States
By Bookmasters